VITAL SIGNS

$+ \quad - \quad \times \quad \div \quad \%$

MATHEMATICS IN EVERYDAY LIFE

VITAL SIGNS

+ − × ÷ %

MATHEMATICS IN EVERYDAY LIFE

TINA LEVY

DIABLO VALLEY COLLEGE

Star
PUBLISHING COMPANY

Star
PUBLISHING COMPANY
P.O. BOX 68
BELMONT, CALIFORNIA 94002
(415) 591-3505

"Dedicated to my mother, Eve, and to the memory of my father, William."

Printed in the United States of America

ISBN: 0-89863-144-0

REVISED PRINTING
9 8 7 6 98 99

TABLE OF CONTENTS

TABLE OF CONTENTS (Continued)

PREFACE

"VITAL SIGNS: MATHEMATICS IN EVERYDAY LIFE" is a basic arithmetic skills workbook intended to teach beginning math students the essential topics needed to continue on to algebra classes and/or basic science classes. It is the revision of the "Basic Skills Workbook" we have been using in our Arithmetic Review classes at Diablo Valley College since 1983. The text is adaptable to both a full semester or quarter class, or to a shorter-term class. At Diablo Valley College we use the text in a short 9-week Arithmetic Review class. It has been class-tested for many semesters.

As a result of many students' and instructors' comments I have expanded both the number of examples in each section as well as the number of exercises for each topic. The text is written in a simple yet explanatory style. I have broken down each concept into its various components so the student is not left guessing what steps to take. I have paid particular attention to making everything clear, giving many examples for each idea. This is especially important for the math-anxious student who often feels that math books assume facts he doesn't know. My work of the past 12 years with both math-anxious students and re-entry women students has given me a sensitivity to what that student needs in a basic text.

I have pored through many sources of data to create interesting problems that will motivate the student. Some of the exercises have strange yet fascinating ideas; other exercises show some of the exciting patterns in arithmetic; while others are the kind of exercises the student will need to master for success in the business or vocational world. I feel that a combination of these types of exercises is unique to this text. The exercises lead students from easy drill of the concepts to more challenging use of the concepts. Wherever possible, I have included application-type problems for the concepts.

Since the text is intended for use in a class that is a *quick* review of basic arithmetic, I assume that the student has a knowledge of the four basic operations with whole numbers. The chapter on Whole Numbers introduces exponents, prime numbers, prime factorization of numbers, least common multiple and greatest common divisor. Many of the returning older students who take the class have never been exposed to these topics.

The chapter on Fractions includes the usual material, but I have also included a section on Ratio and Proportion and one on Complex Fractions. The techniques taught in the Ratio and Proportion section are very useful for problem solving throughout the course. Complex Fractions is a good topic for students to see before algebra.

The chapter on Decimals seemed like the natural spot to introduce Scientific Notation for numbers larger than 1. This exposure is very helpful for students preparing to take a basic science class.

SPECIAL FEATURES

There are 2 practice sets of exercises after each section. Practice Set A has all the answers in the back of the book. Practice Set B contains basically the same kind of exercises as in set A. The problems in set B are of the same difficulty as those in set A but the answers are not given.

This was done to facilitate many different teaching approaches. This text can be used in both a traditional classroom setting or in a laboratory setting.

Use of the calculator is encouraged and introduced in the chapters on Fractions, Decimals and Percents.

Sample exams are included at the end of the chapters on Fractions, Decimals and Percents.

Review exercise sets are given at the end of the chapters on Fractions, Decimals and Percents.

I have included challenging and puzzle-type problems throughout the text. These exercises can be recognized by the * next to them.

The text contains over 2600 exercises. Many of these are the usual drill and practice variety. But a great deal of application-type word problems were also included in as many of the sections as possible. I did this to give students exposure to the type of exercises they will encounter later in classes, on entrance exams as well as in everyday situations. The exercises were drawn from many sources of data.

ACKNOWLEDGEMENTS

I would like to thank the members of the Mathematics Department of Diablo Valley College for their support of my work by their unanimous adoption of this and past versions of the Basic Skills Workbook for use in our Arithmetic program. Many of them were so kind in giving me feedback and many helpful suggestions.

I would like to thank Chuck Wheeler, now retired from the Mathematics Department. It was with him that I first wrote material for use in our Arithmetic Program. I learned a great deal from my collaboration with him. I also want to thank Virginia Johnson and Jeff Mock for their encouragement and support throughout the years.

I want to express great appreciation to Stuart Hoffman, my publisher, for having the foresight to want to publish this text, and for making it turn out so well. It has been a pleasure working with Star Publishing Co.

And most of all, I want to thank my husband, Michael Hreha, for all the encouragement and support he has given me throughout this project. He helped make this happen by always expressing his belief in my abilities.

I hope you will enjoy working through the exercises in the text with your students. If you have any suggestions or comments or other ideas, please feel free to write me. I can be reached at Diablo Valley College, 321 Golf Club Road, Pleasant Hill, California 94523.

HAVE A GOOD CLASS!

Tina Levy

"ON YOUR APPLICATION FOR A LOAN HERE....UNDER DEBTS, WE NEED A LITTLE MORE THAN 'A LITTLE HERE, A LITTLE THERE'".

CHAPTER 1: WHOLE NUMBERS

PLACE VALUE AND READING NUMBERS

• • • etc.	100-trillions	10-trillions	TRILLIONS	100-billions	10-billions	BILLIONS	100-millions	10-millions	MILLIONS	100-thousands	10-thousands	THOUSANDS	hundreds	tens	ONES
• • •															

In our base-10 system each numeral holds the place of a power of 10. We can think of each numeral as naming a specific location. We can read numbers easily if we learn the names of the specific locations. These locations are shown above.

For reading numbers, the important locations are the *third* locations starting after the ONES column and moving left. These names are in capital letters in the diagram above.

Numbers are read in groups of three (hence the commas in large numbers). We start reading from the left and move to the right in the number. The rightmost numeral in any group of 3 numbers separated by commas, determines how that group is read.

For example: 67,315 is read 67 THOUSAND, 315.
In the group 6<u>7</u>, it is the numeral 7 which is in the THOUSANDS place, that names the group.

Another example: 865,708 is read 865 THOUSAND, 708.
In the group 86<u>5</u>, it is the last numeral 5 that names the group since its place is the THOUSANDS place.

Another example: 127,145,009 is read 127 MILLION, 145 THOUSAND, 9
In the group 12<u>7</u>, it is the 7 in the MILLIONS place that names the group.

Then in the group 14<u>5</u>, it is the 5 in the THOUSANDS place that names the group.

LEARNING TO READ VERY LARGE NUMBERS

We may not use very large numbers in our everyday life, but they often appear in newspapers and books, especially as population worldwide is growing, and so is our government deficit.

Just for fun, here is a list of some very large numbers. I have listed the place values that are used in reading the numbers.

1	no places after the numeral	ONES
1,000	3 places after the numeral	THOUSANDS
1,000,000	6 places after the numeral	MILLIONS
1,000,000,000	9 places after the numeral	BILLIONS
1,000,000,000,000	12 places after the numeral	TRILLIONS
1000 . . . 000	15 places after the numeral	QUADRILLIONS
1000 . . . 000	18 places after the numeral	QUINTILLIONS
1000 . . . 000	21 places after the numeral	SEXTILLIONS
1000 . . . 000	24 places after the numeral	SEPTILLIONS
1000 . . . 000	27 places after the numeral	OCTILLIONS
1000 . . . 000	30 places after the numeral	NONILLIONS
1000 . . . 000	33 places after the numeral	DECILLIONS
etc.		

1) Do you see a pattern relating the place value names and the number of places after the numeral? For example, if there are 6 zeros after the numeral, the place is called MILLIONS (think *MONO* ILLIONS).

   If there are 9 zeros after the numeral, that numeral is called BILLIONS (think *BI* ILLIONS).

   Complete the pattern.

2) Let's try to read some numbers together.

   a) 3,000,000,000,000,000 is 3 *QUAD*RILLION (there are 3 × 5 zeros).

   b) 14,000,000,000,000,000,000,000 is 14 *SEX*TILLION (there are 3 × 7 zeros).

   c) 123,000,000,000,000 is 123 *TRI*LLION (there are 3 × 4 zeros).

   d) 415,000,000,897,000 is read as 415 TRILLION, 897 THOUSAND.
   There are 12 places after the 415, therefore it is *tri*llion.
   There are 3 places after the 897, therefore it is thousand.

e) 2,000,548,000,712,000 is read as 2 QUADRILLION, 548 BILLION, 712 THOUSAND.
There are 15 places after the 2, therefore, it is quadrillion.
There are 9 places after the 548, therefore, it is billion.
There are 3 places after the 712, therefore, it is thousand.

f) 2,765,109,890,003 is read as 2 TRILLION, 765 BILLION, 109 MILLION, 890 THOUSAND, 3.
There are 12 places after the 2, therefore it is trillion.
The other place values follow down from trillion.

"GOODNESS GRACIOUS...YOU SAY YOU'RE 104,976,000 SECONDS OLD?!"

PLACE VALUE AND LARGE NUMBERS EXERCISES:

Practice Set A.

1) In the number 27,895,000,000

   a) What place does the 2 hold? 20 billion

   b) What place value does the 7 hold? 7 billion

   c) What place does the 5 hold? 5 million

   d) What place does the 8 hold? 800 million

2) In the number 19,715,436,004,013 name the places indicated by the letters.

   (Arrows under the number mark: a = 1, b = 7, e = 1, c = 4, g = 6, f = 4, d = 3)

   a) 10 trillion

   b) 700 billion

   c) 400 million

   d) ones

   e) 10 billion

   f) 4 thousand

   g) 6 million

Write out in words how you would read the following numbers:

3) 28,400,013 Twenty eight million four hundred thousand thirte[en]

4) 1,000,000,000,001 One trillion one

5) 7,489,116 Seven million four hundred eighty nine one hundred sixteen

6) 40,085,111 Forty million eighty five thousand one hundred eleven

7) 1,111,111,111,111,111,111 One quintrillions

8) 1,010,101,010,101,010 ____________

9) 9,999,909,909,099 ____________

10) 830,000,038 ____________

11) 18,300,000,038,083 ____________

12) 147,095 ____________

Write the symbols for the following numbers. Think carefully about place value.

13) 18 thousand, 34 ____________

14) 678 million, 345 thousand, 11 ____________

15) 2 trillion, 345 thousand, 405 ____________

16) 19 quadrillion, 475 million, 238 thousand ____________

17) 93 billion, 432 million, 780 thousand, 215 ____________

18) 1 quintillion, 5 ____________

19) 13 million, 13 thousand, 13 ____________

20) 999 thousand, 99 ____________

21) 247 thousand, 812 ____________

22) 3 septillion, 4 quintillion, 3 trillion, 7 thousand ____________

23) 3 million, 4 thousand, 899 ____________

24) 4 trillion, 500 thousand, 20 ____________

25) 921 thousand, 99 ____________

26) 45 octillion ____________

*27) one quarter of a billion ____________

PLACE VALUES AND LARGE NUMBERS EXERCISES:

Practice Set B.

1) In the number 43,768,500,000

   a) what place does the 4 hold? tens billions

   b) what place does the 7 hold? hundreds millions

   c) what place does the 5 hold? hundreds thousands

   d) what place does the 6 hold? tens millions

2) In the number 25,463,786,409 name the places indicated by the letters.

   ↑↑ ↑↑↑ ↑ ↑

   a g b f d c e

   a) tens billions

   b) hundreds millions

   c) tens thousands

   d) millions

   e) hundreds

   f) tens millions

   g) billions

Write out in words how you would read the following numbers:

3) 47,600,248 — 47 million 600 thousand 248

4) 2,000,000,000,002 — 2 trillion 2

5) 9,831,246 — 9 million 831 thousand 246

6) 50,075,109 — 50 million 75 thousand 109

7) 22,222,222,222,222 — 22 trillion 222 billion 222 million 222 thousand 222

8) 4,040,040,404,040 — 4 trillion 40 billion 40 million 404 thousand 40

9) 77,707,770,077 — 77 billion 707 million 770 thousand 77

10) 540,000,045 — 540 million 45

11) 98,201,000,002,010 — 98 trillion 201 billion 2 thousand 10

12) 248,763 — 248 thousand 763

Write the symbols for the following numbers. Think carefully about place value.

13) 29 thousand, 56 ______

14) 763 million, 543 thousand, 12 ______

15) 9 trillion, 21 thousand, 208 ______

16) 13 quadrillion, 573 million, 147 thousand ______

17) 98 billion, 507 million, 293 thousand, 181 ______

18) 5 quintillion, 1 ______

19) 35 million, 35 thousand, 35 ______

20) 888 thousand, 88 ______

21) 472 thousand, 753 ______

22) 5 septillion, 4 quintillion, 3 trillion, 1 thousand ______

23) 10 million, 9 thousand, 133 ______

24) 3 trillion, 499 thousand, 19 ______

25) 919 thousand, 89 ______

26) 78 octillion ______

*27) one quarter of a trillion ______

EXPONENTS

When we first learned about multiplication, it was shown to be an operation based on successive addition.

For example, a quantity like 3×5 is the result of doing the addition $3 + 3 + 3 + 3 + 3$ or the addition $5 + 5 + 5$. We either add 5-threes or 3-fives. Multiplication is a shortcut to the successive addition.

We can now define a new operation which is a shortcut to successive multiplication. It is called EXPONENTIATION. When we multiply a number successively a certain number of times, we can abbreviate what we are doing by using what are called "exponents".

For example $3 \times 3 \times 3 \times 3 \times 3$ is indicated as 3^5. We multiplied 5-threes. The 5 is called the "exponent". The 5 tells us how many times to multiply the 3. The 3 is called the "base".

Example 1: In 7^3, the 3 is called the *exponent* (sometimes called the POWER). The 7 is called the *base*.

$7^3 = 7 \times 7 \times 7$ This is called the *expanded* form.

Example 2: Write the expanded form of 8^5.

$8^5 = 8 \times 8 \times 8 \times 8 \times 8$

The multiplication form is called the *expanded* form.

8^5 is called the *exponential* form.

Example 3: Write the exponential form of $3 \times 3 \times 3 \times 3 \times 3 \times 3$.

$3 \times 3 \times 3 \times 3 \times 3 \times 3$ is the expanded form of 3^6.

The exponential form is 3^6.

Example 4: Compute 10^3.

To get numeric answers from exponent problems, we have to first get the expanded form, and then multiply out. There is no shortcut except using a calculator that has the exponent key.

$10^3 = 10 \times 10 \times 10 = 1000$.

The answer is 1000.

Example 5: True or False? Is $2^3 = 6$?

False. Remember that exponents mean successive *multiplication*.

$2^3 = 2 \times 2 \times 2 = 8 \neq 6$ $\quad 2^3 \neq 2 \times 3$

As I mentioned, only calculators provide us with the shortcuts for getting numeric values of exponential problems. Most calculators that are "scientific" or "business-oriented" will have this capacity. Usually they have a key that looks like this:

[y^x] or [x^y]

We are expected to enter the BASE number first, then press the key, then enter the EXPONENT.

If you have a calculator with "reverse Polish" notation, you do the above steps in reverse—you enter the exponent first, then the base.

Example 6: Compute 4^5 on the calculator

We enter 4 [y^x] 5 (=) The answer 1024 will appear in the display.

Example 7: Write the expanded form of 5^4 and compute the answer without a calculator.

$5 \times 5 \times 5 \times 5 = 625$

25 *125* *625*

Example 8: Compute the numeric value of 2^{16} with the help of the calculator.

We press the keys: 2 [y^x] 16 (=) .

The answer will appear in the display *65536*

There are many shortcuts to computations that can be made using exponents. You will see many of them in an algebra class. Here is a simple one to master:

TWO exponential forms with the SAME BASE can be multiplied by simply forming a new exponential form with the same base and the exponents ADDED.

Let's look at the product of $2^3 \times 2^4$.

$$2^3 \times 2^4 = \underbrace{(2 \times 2 \times 2)}_{\text{3-twos}} \times \underbrace{(2 \times 2 \times 2 \times 2)}_{\text{4-twos}} = \underset{\text{is 7-twos}}{\underset{\downarrow}{2^7}}$$

3-twos and 4-twos is 7-twos

Therefore: $2^3 \times 2^4 = 2^7$ (we added the exponents 3 and 4 to get 7)

Numerically this checks out. $2^3 = 2 \times 2 \times 2 = 8$ and $2^4 = 2 \times 2 \times 2 \times 2 = 16$ and $2^7 = 2 \times 2 \times 2 \times 2 \times 2 \times 2 \times 2 = 128$.

Therefore: $2^3 \times 2^4 = 2^7$ since $8 \times 16 = 128$.

Let's try to use this shortcut to find another product of exponentials.

Example 9: Express $3^5 \times 3^6$ as a single exponential.

$$3^5 \times 3^6 = (3 \times 3 \times 3 \times 3 \times 3) \times (3 \times 3 \times 3 \times 3 \times 3 \times 3) = 3^{11}$$

$$3^5 \times 3^6 = 3^{5+6} = 3^{11}$$

Example 10: Express $8^4 \times 8^2$ as a single exponential.

$$8^4 \times 8^2 = 8^{4+2} = 8^6$$

EXPONENTS EXERCISES:

Practice Set A

Write the exponential form of the number whose *base* and *exponent* are given:

1) base 3, exponent 5 2) base 6, exponent 3 3) base 8, exponent 9

4) base 9, exponent 8 5) base 10, exponent 4 *6) base 3^2, exponent 5

Write the expanded form of each of the following exponentials, and compute the answer without a calculator:

7) 2^7 8) 3^2 9) 4^3 10) 5^3 11) 7^2

12) 8^3 13) 11^2 14) 9^3 15) 2^5 16) 3^5

17) Write the expanded form of 10^3, 10^4, 10^5, and 10^6. Compute these answers. What is the pattern with these numbers?

18) Compute the values of the following exponentials:

2^8, 2^7, 2^6, 2^5, 2^4, 2^3, 2^2, and 2^1.

What is the pattern in the numeric values of these powers as the exponents decrease in value?

Guess what the value of 2^0 would be.

Do the following computations *with a calculator*:

19) 9^7 20) 8^8 21) 11^5 22) 5^{11} 23) 2^{25}

24) 101^3 25) 15^6 26) 2001^2 27) 89^3 28) 1986^2

Use the shortcut to express each of the following as a single exponential:

29) $4^2 \times 4^3$ 30) $3^6 \times 3^7$ 31) $2^5 \times 2^8$ 32) $10^3 \times 10^4$

33) $12^9 \times 12^{11}$ 34) $2^{151} \times 2^{174}$ *35) $9^{\square} \times 9^{\Delta}$

36) How would you interpret 3^{3^3}?

* * 37) Try to describe how large $10^{10^{10}}$ is and how it would be written.

38) Every person born has a mother and a father. Let us say they are your ancestors 1 generation back. So you have 2 ancestors in 1 generation back. Each of them also had a mother and a father. Those people would be your ancestors 2 generations back. If we continue this reasoning, we can find out how many ancestors any number of generations back:

   a) How many ancestors do you have 2 generations back?
   b) How many ancestors do you have 3 generations back?
   c) 4 generations back?
   d) 10 generations back?

39) Monday, Corina hears a rumor. On Tuesday she tells it to two other people. That makes 3 people in total that know the rumor on Tuesday. On Wednesday, each person who knew the rumor the day before tells it to 2 other people. And so the rumor begins to spread.

   a) How many people in total know the rumor on Friday?
   b) One week after the rumor started, how many people know about it?

* *40) There is a lovely pond, and on the pond there is one beautiful lily to start. The number of lilies double every 24 hours. In 59 days the pond is half-covered. How many days will it take for the lilies to cover the pond entirely?

EXPONENTS EXERCISES:

Practice Set B

Write the exponential form of the number whose *base* and *exponent* are given:

1) base 4, exponent 3 2) base 8, exponent 4 3) base 7, exponent 6

4) base 6, exponent 7 5) base 10, exponent 6 6) base 2^3, exponent 4

Write the expanded form of each of the following exponentials, and compute the numeric value without a calculator:

7) 3^6 8) 4^2 9) 5^3 10) 7^3 11) 8^2

12) 6^3 13) 12^2 14) 11^3 15) 2^8 16) 9^2

Do the following computations with the aid of a calculator:

17) 3^{12} 18) 11^7 19) 2^{20} 20) 5^{10} 21) 10^5

22) 7^8 23) 8^7 24) 9^8 25) 13^6

Use the shortcut to express each of the following as a single exponential:

26) $7^4 \times 7^5$ 27) $3^6 \times 3^4$ 28) $2^{10} \times 2^9$ 29) $10^3 \times 10^5$

30) $11^4 \times 11^{11}$ 31) $12^{15} \times 12^{25}$ 32) $2^{256} \times 2^{145}$ *33) $5^{⬠} \times 5^{⬡}$

34) The number 1 followed by 100 zeros is a very large number. It was given the name "googol" by the young nephew of the American mathematician Edward Kasner. How would you write the number as a single power of 10?

Try to write the number as a power of 10 whose exponent is also a power of 10.

SHORTCUTS TO MENTAL ARITHMETIC

Multiplying By Powers of 10:

If we want to multiply any whole number by any power of 10, it is a simple matter of just counting the number of zeros in the power of 10, and putting those zeros on the end of the whole number. Then insert the commas in the appropriate place.

Example 1: $34 \times 100 = 34\underline{00} = 3{,}400$

$156 \times 10000 = 156\underline{0000} = 1{,}560{,}000$

Powers of 10:

$10^1 = 10$
$10^2 = 100$
$10^3 = 1000$
$10^4 = 10000$
$10^5 = 100000$
$10^6 = 1000000$
etc.

Multiplying by 5 or 25:

If we want to multiply any number by 5 or 25, it is a simple matter of just halving the number once or twice, and then proceeding as if we were multiplying by 10 or 100.

To multiply a number by 5, we halve the number (divide by 2) and then multiply the result by 10.

Example 2: $256 \times 5 = 256 \times \frac{10}{2} = \frac{256 \times 10}{2} = \frac{256}{2} \times 10 = 128 \times 10 = 1280$

Example 3: Multiply 284 by 5.

$2\overline{)284}$ = 142 then $142 \times 10 = 1420$

You will note that this shortcut works with numbers that are EVEN. Later, when we learn about decimals, you can apply it to odd numbers.

To multiply a number by 25, we halve the number once, then that result gets halved again. We then multiply by 100.

Example 4: $1024 \times 25 = 1024 \times \frac{100}{4} = \frac{1024 \times 100}{4} = \frac{1024 \times 100}{2 \times 2} = \frac{1024}{2} \times \frac{100}{2} =$

$512 \times \frac{100}{2} = \frac{512}{2} \times 100 = 256 \times 100 = 25600.$

The quick result could have been gotten by taking half of 1024 which is 512, and then taking half of 512 which is 256. Then we multiply 256 by 100 which gives our result of 25,600.

Example 5: Multiply 432 by 25.

$2\overline{)432}$ = 216 then $2\overline{)216}$ = 108 then $108 \times 100 = 10{,}800$

Example 6: Multiply 12,816 by 25.

$2\overline{)12816}$ = 6408 then $2\overline{)6408}$ = 3204 then $3204 \times 100 = 320{,}400$.

Example 7: Multiply 2460×25

$2\overline{)2460}$ = 1230 then $2\overline{)1230}$ = 615 and $615 \times 100 = 61{,}500$

You will notice that the shortcut is applied to numbers that are divisible by 4. Later, when we learn about decimals, you can use the shortcut on any number.

MULTIPLYING TWO NUMBERS BETWEEN 10 AND 20.

We will try a shortcut to multiply 17×19.

The steps are:

a) Add the one number and the *ones digit* of the other number — $17 + 9 = 26$

b) Multiply the result by 10. — $26 \times 10 = 260$

c) Multiply both the ones digits — $7 \times 9 = 63$

d) Add the results of steps b and c to get the answer. — $260 + 63 = \mathit{323}$

Example 8: Multiply 18×15.

a) Add one number and the ones digit of the other. $15 + 8 = 23$

b) Multiply the result by 10. $23 \times 10 = 230$

c) Multiply the ones digits of both numbers. $8 \times 5 = 40$

d) Add the results of steps b and c to get the answer. $230 + 40 = 270$

The answer is 270.

SHORTCUTS TO MENTAL ARITHMETIC EXERCISES:

Practice Set A

Multiply each of the following numbers by 5:

1) 8192	2) 254	3) 144	4) 2008	5) 10,456
6) 8762	7) 41,222	8) 28	9) 3330	10) 68,004
11) 192	12) 46	13) 9766	14) 876	15) 998

Multiply each of the following numbers by 25:

16) 8192	17) 160,116	18) 48	19) 128	20) 964
21) 4020	22) 21,284	23) 788	24) 79,996	25) 85,956

Multiply the following two numbers by the shortcut in this section:

26) 12×16	27) 18×19	28) 19×19	29) 14×16	30) 13×15
31) 15×17	32) 16×16	33) 11×17	34) 12×18	35) 11×19

SHORTCUTS TO MENTAL ARITHMETIC EXERCISES:

Practice Set B

Multiply each of the following numbers by 5:

1) 9182 2) 456 3) 448 4) 8002 5) 18,656

6) 1914 7) 46 8) 98 9) 154 10) 3460

11) 78,122 12) 2002 13) 5516 14) 18 15) 1998

Multiply each of the following numbers by 25:

16) 9812 17) 481,156 18) 428 19) 68 20) 2308

21) 4200 22) 79,600 23) 87,656 24) 288 25) 4004

Multiply the following two numbers by the shortcut in this section:

26) 13×16 27) 15×18 28) 18×18

29) 19×12 30) 11×16 31) 14×18

32) 14×11

RULES FOR DIVISIBILITY

Rule for Dividing by 2:

A number is divisible by 2 if it is an EVEN number; i.e., it ends in the digits 0, 2, 4, 6, 8.

Rule for Dividing by 3:

A number is divisible by 3 if the sum of the digits of the number is divisible by 3.

Example: 963 is divisible by 3 since $9 + 6 + 3 = 18$, and 3 divides 18 evenly.

Rule for Dividing by 4:

A number is divisible by 4 if it is EVEN and the last two digits are divisible by 4.

Example: 2048 is divisible by 4 since it is even and the last two digits, 48, are divisible by 4.

Rule for Dividing by 5:

A number is divisible by 5 if it ends in a 5 or a 0.

Example: 2,109,478,345 is divisible by 5 since the last digit is a 5.

Rule for Dividing by 6:

A number is divisible by 6 if it is BOTH even and such that the sum of its digits is divisible by 3. Really we are saying that the number is divisible by BOTH 2 and 3.

Example: 2142 is divisible by 6. It is even, since it ends in a 2, and the sum of its digits $2 + 1 + 4 + 2 = 9$, which is divisible by 3.

Rule for Dividing by 8:

A number is divisible by 8 if the last 3 digits of the number are divisible by 8. If a number has less than 3 digits, then we just divide it by 8.

Example: 8192 is divisible by 8 since the last three digits, 192, can be evenly divided by 8. $(192 \div 8 = 24)$

Rule for Dividing by 9:

A number is divisible by 9 if the sum of its digits is divisible by 9.

Example: 91,845 is divisible by 9 since the sum of its digits is 27 (9 + 1 + 8 + 4 + 5 = 27), and 27 is divisible by 9.

We can continue adding digits successively, and if we get a 9, then the number is divisible by 9. This is a shortcut. So, in the example above, since 2 + 7 = 9, it confirms the result.

Rule for Dividing by 10:

A number is divisible by 10 if it ends in a 0.

Example: 45,670 is divisible by 10 since it ends in a 0.

The rules for 7, 11 and 13 are too complicated for us now.

We will try some of these rules on various numbers to decide which of the above numbers divide evenly into them.

Example 1: List all the numbers that divide 189,888.

Since the number ends in 8, it is divisible by 2.

Since the sum of the digits is 42 (1 + 8 + 9 + 8 + 8 + 8 = 42) and 42 is divisible by 3, the number is divisible by 3.

Since the last two digits are 88, and 88 is divisible by 4, the number is divisible by 4.

Since the number is BOTH even and divisible by 3, the number is also divisible by 6.

Since the last three digits, 888, are divisible by 8, then the number is divisible by 8.

The number is NOT divisible by 9 since the sum of the digits, 42, is not divisible by 9. Using the shortcut, since 4 + 2 = 6, not 9, it is not divisible by 9.

RULES FOR DIVISIBILITY EXERCISES:

Practice Set A

List all the numbers 2, 3, 4, 5, 6, 8, 9, or 10, that divide evenly into the following numbers. You can always check the answers on the calculator, but try them first without the calculator.

1) 91,845	2) 2048	3) 178,605	4) 2,300,103	5) 123,456
6) 111,111,111	7) 213	8) 921,912	9) 912,921	10) 10,101
11) 2934	12) 918,306	13) 38,318,896	14) 344,870,064	15) 243,450
16) 415	17) 2544	18) 1,000,011	19) 53,078,886	20) 477,709,974

21) Why is it that the product of any three consecutive numbers is always divisible by 6?

22) What values of d will make the following numbers divisible by 9?

a. $821d$ b. $82d1$ c. $43{,}250{,}0d4$

23) What values of d will make the following numbers divisible by 6?

a. $821d$ b. $114{,}3d2$ c. $43{,}250{,}0d0$

RULES FOR DIVISIBILITY EXERCISES:

Practice Set B

List all the numbers 2, 3, 4, 5, 6, 8, 9, or 10 that divide evenly into the following numbers:

1) 61,776
2) 4096
3) 101,000,001
4) 51,111
5) 1,234,242
6) 312
7) 25,536
8) 432
9) 1287
10) 6552
11) 5110
12) 84,455
13) 54,162,162,054

14) What values of d will make the following numbers divisible by 9?

a. 183d b. 831d1 c. 52,788,0dd

15) What values of d will make the following numbers divisible by 6?

a. 183d b. 83d1 c. 52,788,0dd

PRIME FACTORIZATION OF WHOLE NUMBERS

We can think of all whole numbers as if they were molecules formed by irreducible parts, like atoms, that are all connected by the operation of multiplication. These irreducible parts, which I compare to atoms, are the building blocks of all numbers. They are called PRIME NUMBERS.

All numbers can be expressed as the product of these *prime numbers*.

Example: The number 6 can be thought of as a molecule consisting of two atoms combined through multiplication. The two atoms are the prime numbers 2 and 3. They are the building blocks that form the number 6.

> A PRIME NUMBER is a number with exactly two divisors, itself and 1. It cannot be formed by multiplying any other numbers. It is like an atom in that it is a basic unit that cannot be decomposed.

Example: The number 5 is a prime number, but the number 9 is not. The number 9 can be formed by 9×1 or by 3×3. The number 15 is NOT prime since it can be formed by 1×15 and by 3×5.

Any number that is not prime is called a COMPOSITE number. Both the numbers 9 and 15 are composite.

Here are the first 10 prime numbers:

2, 3, 5, 7, 11, 13, 17, 19, 23, 29, . . .

There are infinitely many prime numbers. There is a systematic way of generating all the primes up to a certain number. The method comes from ancient times and is called the Sieve of Eratosthenes.

Let us suppose that we want to generate all the prime numbers between 1 and 100. We make a grid of all the numbers from 1 to 100 by listing them in rows of 10 like this:

1	2	3	4	5	6	7	8	9	10
11	12	13	14	15	16	17	18	19	20
21	22	23	24	25	26	27	28	29	30
31	32	33	34	35	36	37	38	39	40
41	42	43	44	45	46	47	48	49	50
51	52	53	54	55	56	57	58	59	60
61	62	63	64	65	66	67	68	69	70
71	72	73	74	75	76	77	78	79	80
81	82	83	84	85	86	87	88	89	90
91	92	93	94	95	96	97	98	99	100

First we cross off the 1. (1 is not considered a prime number since its only divisor is itself).

Now we look at the next number in the grid, which is 2. We circle it.

No other numbers multiply to make 2 except 2 and 1, so it is prime. We now proceed to cross off any number that 2 evenly divides. We go through the grid, row by row and entry by entry, until we are at the end.

Now we go back and see which number is next and hasn't been crossed out. It is the number 3, so we circle it since it is prime, and proceed to cross out all the numbers that 3 divides into evenly. We continue in this manner row by row and entry by entry until we reach the end of the grid. We then go back to the beginning and circle the next number NOT crossed out. That number is the next prime.

The grid will look like this after the first two rounds:

~~1~~	②	③	~~4~~	⑤	~~6~~	7	~~8~~	~~9~~	~~10~~
11	~~12~~	13	~~14~~	~~15~~	~~16~~	17	~~18~~	19	~~20~~
~~21~~	~~22~~	23	~~24~~	25	~~26~~	~~27~~	~~28~~	29	~~30~~
31	~~32~~	~~33~~	~~34~~	35	~~36~~	37	~~38~~	~~39~~	~~40~~
41	~~42~~	43	~~44~~	~~45~~	~~46~~	47	~~48~~	49	~~50~~
~~51~~	~~52~~	53	~~54~~	55	~~56~~	~~57~~	~~58~~	59	~~60~~
61	~~62~~	~~63~~	~~64~~	65	~~66~~	67	~~68~~	~~69~~	~~70~~
71	~~72~~	73	~~74~~	~~75~~	~~76~~	77	~~78~~	79	~~80~~
~~81~~	~~82~~	83	~~84~~	85	~~86~~	~~87~~	~~88~~	89	~~90~~
91	~~92~~	~~93~~	~~94~~	95	~~96~~	97	~~98~~	~~99~~	~~100~~

We continue the process through the grid crossing out all the multiples of the new prime number. When we get through the grid, we go back to the next prime, and so on, and so on and so on.

It will be obvious when we are finished. All the numbers of the grid will be either crossed out or circled.

QUESTION: Why is it called a "sieve"?

ANSWER: Only the prime numbers will "fall through" after we eliminate the various multiples of primes.

Now that we have a way of generating *prime numbers,* we can generate the PRIME FACTORIZATIONS of any whole number. Each number can be expressed as a unique product of prime numbers.

There are several ways to accomplish this.

Example 1: Find the prime factors of 100.

$$\begin{array}{rcccccc} 100 = & 2 & \times & & 50 & & \\ & \downarrow & & & \swarrow\searrow & & \\ = & 2 & \times & 25 & \times & 2 & \\ & \downarrow & & \swarrow\searrow & & \searrow & \\ = & 2 & \times & 5 \times 5 & \times & 2 & \end{array}$$

This is called a "factor tree". We find any product of 100 and then continue to break each factor down into its components until we are left with only prime numbers as factors.

The factor tree method could become very long if the number is large. There is a simpler method that involves successive division.

We try the first prime number that divides evenly into 100. It is 2.

$$\begin{array}{r} 2\overline{)100} \\ 50 \end{array} \qquad (100 \div 2 = 50)$$

We continue to divide 2 into each result, until it doesn't divide anymore.

$$\begin{array}{r} 2\overline{)50} \\ 25 \end{array} \qquad (50 \div 2 = 25)$$

Then we go to the next prime number which divides evenly into the last result.

$$\begin{array}{r} 5\overline{)25} \\ 5 \end{array} \qquad (25 \div 5 = 5)$$

We continue in this way until we eventually get 1.

$$\begin{array}{r} 5\overline{)5} \\ 1 \end{array} \qquad (5 \div 5 = 1)$$

Let's put this all together in one continuous sequence.

$$\begin{array}{r} 2\overline{)100} \\ 2\overline{)50} \\ 5\overline{)25} \\ 5\overline{)5} \\ 1 \end{array}$$

We can then read off all the factors: 2, 2, 5, 5.

The prime factors of 100 are $2 \times 2 \times 5 \times 5 = 2^2 \times 5^2$

Example 2: Find the prime factors of 120.

$120 = 2 \times 2 \times 2 \times 3 \times 5$

$$\begin{array}{r} 2\overline{)120} \\ 2\overline{)60} \\ 2\overline{)30} \\ 3\overline{)15} \\ 5\overline{)5} \\ 1 \end{array}$$

The prime factorization of $120 = 2 \times 2 \times 2 \times 3 \times 5 = 2^3 \times 3 \times 5$

Example 3: Find the prime factorization of 18.

$$\begin{array}{r} 2\overline{)18} \\ 3\overline{)9} \\ 3\overline{)3} \\ 1 \end{array} \qquad 18 = 2 \times 3^2$$

Example 4: Find the prime factorization of 1001.

$$\begin{array}{r} 7\overline{)1001} \\ 11\overline{)143} \\ 13\overline{)13} \\ 1 \end{array} \qquad 1001 = 7 \times 11 \times 13$$

This problem is a bit tricky. The first prime number that worked was 7. You have to be patient and keep trying primes until you find the first divisor.

PRIME FACTORIZATION OF WHOLE NUMBERS EXERCISES:

Practice Set A

1) What is a fast way of crossing out all the multiples of a prime in the grid?

2) Using the grid, list all the primes less than 100.

3) Give several ways that 100 is *not* a prime.

4) Why is NO even number, other than 2, a prime number?

Find the prime factorization of each of the following numbers.

5) 12	6) 40	7) 36	8) 50	9) 81
10) 242	11) 65	12) 459	13) 111	14) 2001
15) 360	16) 1000	17) 48	18) 128	19) 92
20) 60	21) 216	22) 336	23) 125	24) 27

*25) As a real challenge, use the sieve method to find all the primes less than 1000.

PRIME FACTORIZATION OF WHOLE NUMBERS EXERCISES:

Practice Set B

Find the prime factorization of each of the following numbers:

1) 153	2) 610	3) 42	4) 500	5) 75
6) 240	7) 150	8) 108	9) 285	10) 31
11) 300	12) 123	13) 729	14) 294	15) 561
16) 81	17) 105	18) 70	19) 126	*20) 2717

A PICTORIAL PATTERN USING PRIME FACTORIZATIONS (Optional Topic)

Once we have identified prime numbers, we can get the prime factorizations of any number. It is not always easy, but we can always do it after some testing.

We can now make diagrams of some of these numbers using their prime factorizations. We will call these "lattice diagrams."

Let's start all our numbers with the factor 1, which is neither prime nor composite.

If a number is made up of only one kind of prime factor, then it can be represented by a LINE (or a series of line segments).

Example 1: Diagram the number 5.

Example 2: Diagram the number 25.

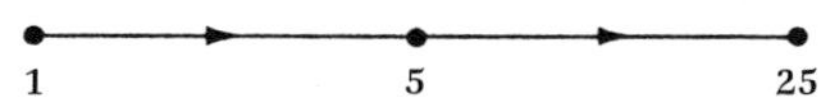

Each successive factor of 5 extends the line by a SEGMENT. Since 25 is made up ot TWO factors of 5, there are TWO line segments in the diagram. As we move to the right, we keep multiplying by 5.

If a number has two different prime factors, then it can be represented as a square (or a series of squares). The 2 kinds of prime factors give us a diagram in 2 dimensions.

Example 3: Diagram the number 6.

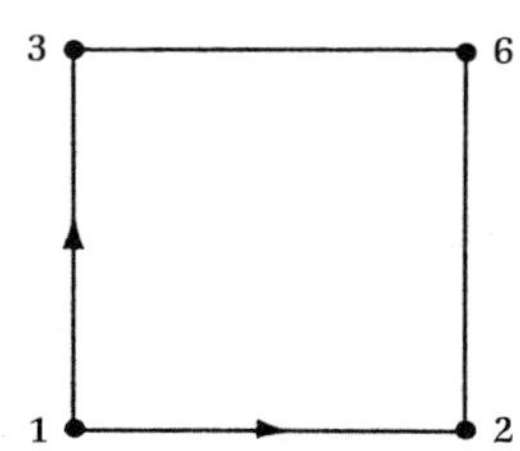

$6 = 2 \times 3$

Each of the two factors is connected by a line to the factor 1. Since they are different factors, they go in different directions. Going to the right, we are multiplying by 2. Going up, we are multiplying by 3.

Example 4: Diagram the number 12.

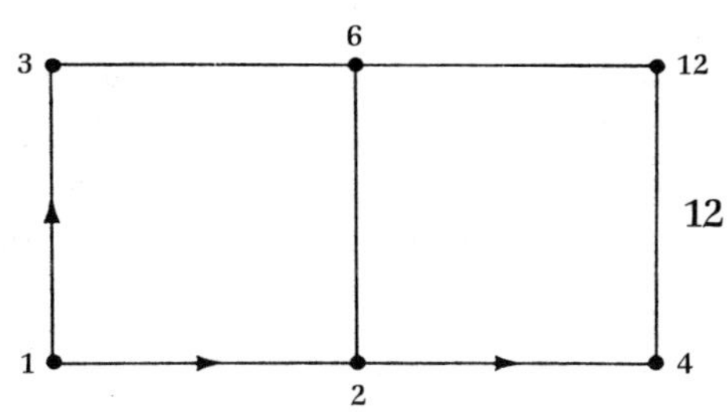

$12 = 2^2 \times 3$

As we move to the right, we are multiplying each number by 2. Since there are TWO factors of 2, we are moving right TWICE. As we move up, we are multiplying by 3. Since there is only ONE factor of three, we are moving up only ONCE.

Example 5: Diagram the number 18:

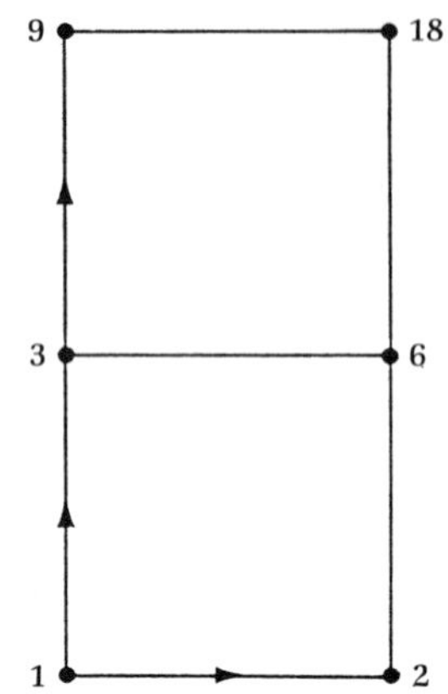

$$18 = 2 \times 3^2$$

Since there are TWO factors of 3, we move up TWICE. As we move up, each number is being multiplied by 3. Since there is only ONE factor of 2, we move right only ONCE.

When we have one prime factor in a number, we get a diagram that involves line segments. The diagram is one-dimensional.

When we have two different prime factors in a number, we get a diagram that involves squares. The diagram is two-dimensional.

QUESTION: When we have three different prime factors in a number, what kind of figure will the diagram of the number involve?

ANSWER: Right! We will get a CUBE (or a series of cubes).

Example 6: Diagram the number 30.

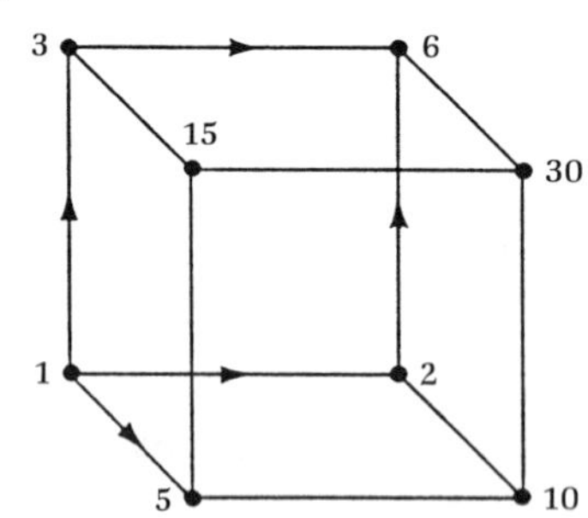

$$30 = 2 \times 3 \times 5$$

QUESTION: What are the three different directions?

QUESTION: What happens as we move in each of the three directions?

PICTORIAL PATTERNS USING PRIME FACTORIZATIONS EXERCISES:

Practice Set A

Diagram each of the following numbers:

1) 13	2) 9	3) 22	4) 45	5) 75
6) 24	7) 100	8) 42	9) 54	10) 60

Each of the following lattice diagrams represents a particular number. Fill in the missing numbers for each diagram and tell what number is represented by each picture.

11)

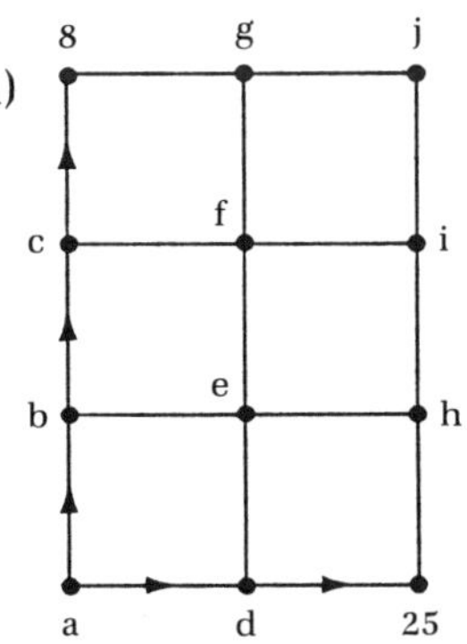

12)

13)

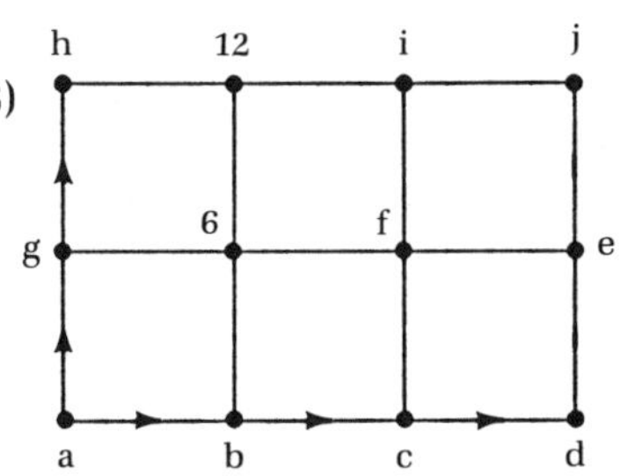

14)

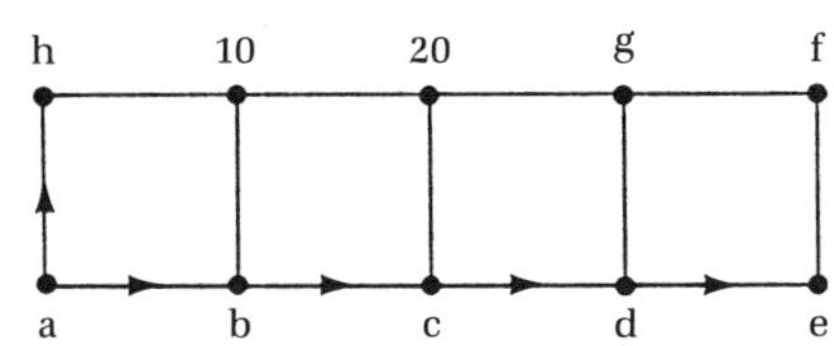

PICTORIAL PATTERNS USING PRIME FACTORIZATIONS EXERCISES:

Practice Set B

Diagram each of the following numbers:

1) 11	2) 49	3) 39	4) 63	5) 175
6) 250	7) 36	8) 105	9) 135	10) 120

Each of the following lattice diagrams represents a particular number. Fill in the missing numbers for each diagram and tell what number is represented by each picture.

11) 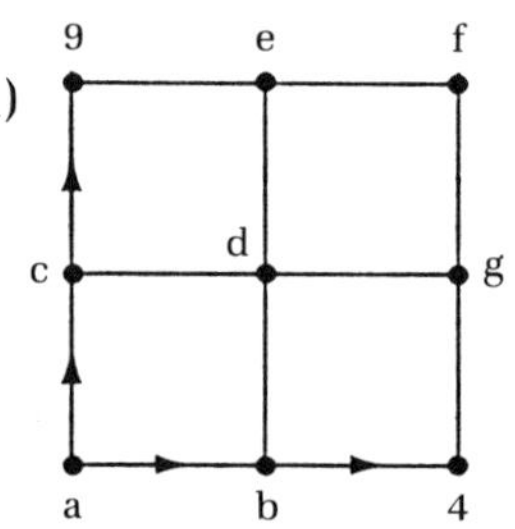

12) 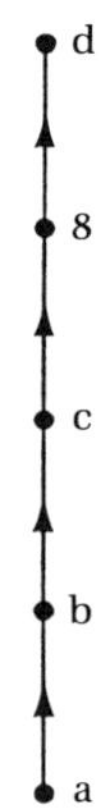

13) 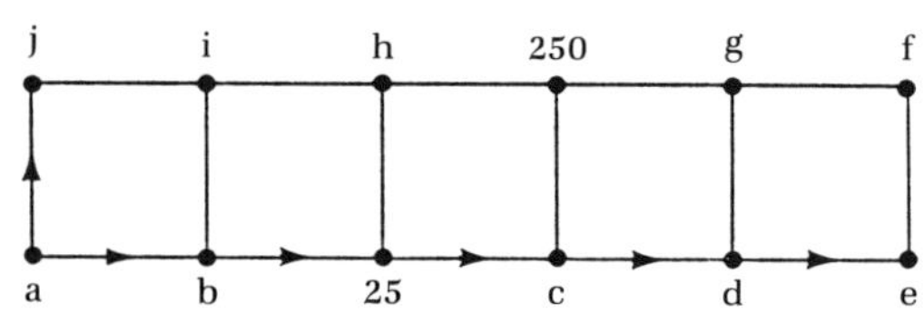

14) 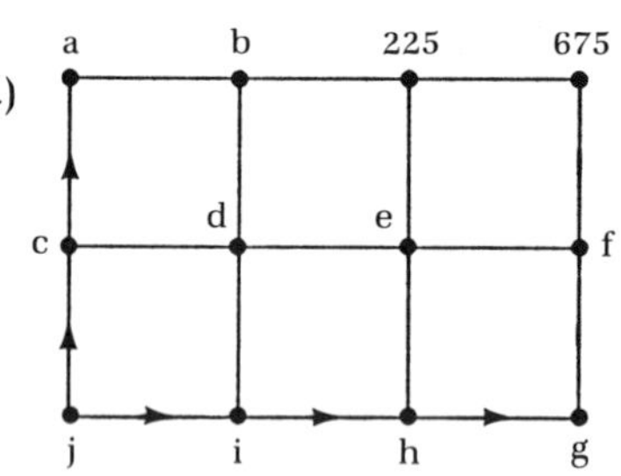

GREATEST COMMON DIVISOR

When we work with fractions it will be necessary to compare two (or more) numbers to determine the LARGEST number that divides evenly into BOTH numbers. We do this when reducing fractions to lowest terms.

The largest number that divides evenly into all the numbers being considered is called the GREATEST COMMON DIVISOR. It is abbreviated as g.c.d.

Example 1: Find the g.c.d. of 10 and 25.

Looking at the two numbers and thinking of what numbers divide them evenly we come up with 5. By inspection we can see that 5 is the largest number that divides both.

If we had to rely on inspection, we would not always be able to do every problem. However, there is a systematic way of finding the g.c.d. of any numbers using their prime factorizations.

TO FIND THE G.C.D. OF ANY NUMBERS WE LIST THEIR PRIME FACTORIZATIONS AND THEN CHOOSE ALL THE FACTORS THAT ARE *COMMON* TO ALL THE NUMBERS BEING CONSIDERED.

Example 2: Find the g.c.d. of 10 and 25 using their prime factors.

$$\begin{aligned} 10 &= 2 \times 5 \\ 25 &= 5 \times 5 \\ \hline \text{g.c.d.} &= 5 \end{aligned}$$

Example 3: Find the g.c.d. of 18 and 30.

$$\begin{aligned} 18 &= 2 \times 3 \times 3 \\ 30 &= 2 \times 3 \times 5 \\ \hline \text{g.c.d.} &= 2 \times 3 = 6 \end{aligned}$$

Example 4: Find the g.c.d. of 27 and 45.

$$\begin{aligned} 27 &= 3^3 \\ 45 &= 3^2 \times 5 \\ \hline \text{g.c.d.} &= 3^2 = 9 \end{aligned}$$

(Exponents make it easy to see that both numbers have only two 3s in common. We choose the smallest exponent of each prime factor.)

Example 5: Find the g.c.d. of 15, 27 and 22.

$$\begin{array}{l} 15 = 3 \times 5 \\ 27 = 3^3 \\ \underline{22 = 2 \times 11} \\ \text{g.c.d.} = 1 \end{array}$$

There are NO factors in common with *all* the numbers. For that reason we say the g.c.d. is 1.

Numbers whose g.c.d. is 1 are called RELATIVELY PRIME.

GREATEST COMMON DIVISOR EXERCISES:

Practice Set A

Find the g.c.d. of the following numbers:

1) 12 and 30
2) 21 and 45
3) 22 and 55
4) 20 and 90
5) 340 and 100
6) 126 and 198
7) 27, 87 and 48
8) 10, 20 and 60
9) 10, 100 and 1000
10) 15, 14 and 10
11) 9, 36 and 15
12) 90, 12 and 54
13) 144 and 216
14) 45, 32 and 54
15) 28 and 77
16) 1001 and 121
17) 125 and 175
18) $2^3 \times 3^2 \times 5^7$, $2^7 \times 3^2 \times 7^5$, $2^5 \times 3 \times 5 \times 7$

GREATEST COMMON DIVISOR EXERCISES:

Practice Set B

Find the g.c.d. of the following numbers:

1) 18 and 32

2) 36 and 52

3) 34 and 51

4) 30 and 70

5) 225 and 375

6) 128 and 112

7) 39, 81 and 45

8) 15, 30 and 90

9) 6, 36 and 216

10) 16, 15 and 20

11) 8, 48 and 28

12) 169 and 1001

13) 196 and 84

14) 15, 100 and 75

15) $2^4 \times 3^4 \times 7$, $2 \times 5^2 \times 7^2$, $2 \times 3^2 \times 5^2 \times 7$

LEAST COMMON MULTIPLE

When we work with fractions, we will need to find the smallest number that can be divided BY a given set of numbers. It is at least as large as the largest number in the set. This smallest number is called the LEAST COMMON MULTIPLE. We abbreviate it as l.c.m.

Example 1: Find the l.c.m. of 6 and 9

We begin by generating all the multiples of 6.
They are 6, 12, 18, 24, 30, 36, 42, 48, . . . etc.

We then generate all the multiples of 9.
They are 9, 18, 27, 36, 45, 54, 63, 72, 81, . . . etc.

We now choose the SMALLEST of the numbers COMMON to both sets of *multiples*.

That number is 18. It is the smallest number that can be divided by both the 6 and the 9.

If we had to use this method of multiples to find every least common multiple of given numbers it would take too long. There is an easier way using the prime factorizations of the given numbers.

Example 2: Find the l.c.m. of 6 and 9.

$$\begin{array}{r} 6 = 2 \times 3 \\ \underline{9 = 3 \times 3 \qquad} \\ \text{l.c.m.} = 2 \times 3 \times 3 = 18 \end{array}$$

The l.c.m. must have one of EACH prime that is in the factorizations of each number. It must have that prime the MOST number of times it occurs in any one number.

We used the 2 only once since it occurred only once in 6. But we had to use the 3 twice since it occurs in 9 twice.

Example 3: Find the l.c.m. of 8, 12 and 6.

$$\begin{array}{r} 8 = 2 \times 2 \times 2 \\ 12 = 2 \times 2 \times 3 \\ \underline{6 = 2 \times 3 \qquad\qquad} \\ \text{l.c.m.} = 2 \times 2 \times 2 \times 3 = 24 \end{array}$$

We have to use the 2 three times since it occurs in the 8 three times. We use the 3 once since it occurs only once in any number.

In effect, we choose the *largest* exponent of each prime factor.

$$8 = 2^3$$
$$12 = 2^2 \times 3^1$$
$$6 = 2 \times 3^1$$

$$\text{So l.c.m.} = 2^3 \times 3^1$$

Example 4: Find the l.c.m. of 10, 21 and 11.

$$10 = 2 \times 5$$
$$21 = 3 \times 7$$
$$11 = 11$$
$$\text{l.c.m.} = 2 \times 3 \times 5 \times 7 \times 11 = 2310$$

We had to use each factor exactly once since no prime occurs more than once in any of the numbers.

Example 5: Find the l.c.m. of these two numbers, n and m.

$$n = 2^3 \times 3^2 \times 5^1$$
$$m = 2^2 \times 3^4 \times 7^1$$

$$\text{l.c.m.} = 2^3 \times 3^4 \times 5^1 \times 7^1$$

We used the prime 2 *three* times since it occurs in number "n" three times.

The most number of times that 3 appears is in number "m"—four times. So we have four 3's in the l.c.m.

The factors 5 and 7 each appear once.

LEAST COMMON MULTIPLE EXERCISES:

Practice Set A

Find the least common multiples of the following numbers:

1) 10 and 12
2) 4, 6 and 5
3) 3, 2 and 5
4) 100 and 25
5) 18 and 21
6) 12, 15 and 20
7) 27, 18, 36 and 4
8) 6, 15 and 10
9) 3, 4, 8 and 6
10) 10, 100 and 1000
11) 60 and 72
12) 8, 12 and 9
13) 12 and 16
14) 10 and 8
15) 24 and 15
16) 35 and 315
17) 48 and 192
18) 65 and 91
19) 11, 13 and 7
20) 485 and 291

Here are some fun applications of LCM:

21) Two dials, one with 6 numerals and the other with 8 numerals, equally spaced, show 1 and 1 at the start. If each one advances one position per minute, when will they again show 1 and 1 at the *same* time?

22) The Mertons live on a block that has two very unusual dogs. These dogs are perfect angels during the daylight hours, but they begin to bark as soon as the sun sets and they do so very regularly. They only bark for a second but they do this all night long. The Curtin dog barks once every 21 minutes. The Burton dog barks once every 15 minutes.

Assuming that the two dogs start at exactly the same moment, at 6:40 PM one evening, what time next will they BOTH be barking?

23) The Girl Scouts Association is packaging their cookies again this year. A certain number of cookie boxes can be packed into cartons holding 25 boxes, with each carton completely filled. It is also possible to pack the same number of cookie boxes in cartons holding 30 boxes, and the cartons would be completely filled. What is the smallest number of Girl Scout cookie boxes that we could have and yet satisfy both conditions?

24) One satellite orbits planet Zirconia once every 48 hours. Another satellite orbits the planet once every 60 hours. If they are both above the same point on planet Zirconia at the same time, when will they be there next?

25) A bus travels from San Francisco to San Jose every 24 minutes. Another bus travels from Redwood City to San Jose every 40 minutes. The buses from San Francisco and Redwood City just arrived at the same time. How long will it take before this happens again?

26) If one car gets 18 miles to the gallon, and another car gets 16 miles to the gallon, what is the *smallest* number of gallons of gasoline they should have in their tanks to go exactly the same distance?

LEAST COMMON MULTIPLE EXERCISES:

Practice Set B

Find the least common multiples of the following numbers:

1) 12 and 20
2) 5, 8 and 10
3) 2, 5 and 7
4) 256 and 32
5) 21 and 24
6) 12, 18 and 30
7) 28, 21, 42 and 6
8) 9, 12 and 10
9) 5, 6, 12 and 10
10) 11, 121 and 1331
11) 100 and 1000
12) 18 and 16
13) 75 and 60
14) 9, 10 and 11
15) 32 and 18
16) 225 and 15
17) 213 and 639
18) 119 and 153
19) 42 and 48
20) 90, 12 and 24

Here are some fun examples of LCM:

21) Two dials, one with 5 numerals and the other with 7 numerals, equally spaced, show 1 and 1 at the start. If each one advances one position per minute, when will they again show 1 and 1 at the *same* time?

22) A train arrives in London from Manchester once every hour. Another train arrives in London from Brighton once every 1 hour and 15 minutes. If both trains are at the same station in London at 3:45 pm, when will they both be there together again? (HINT: change hours to minutes)

23) Your beautiful rhododendron bush blooms once every 4 days in the spring. Your special variety rosebush produces exquisite blooms once every 7 days in Spring and Summer. But your every rare and exotic orchid will produce 1 bloom once every 13 days during the month of May only. Today is May 10 and all three are blooming at once. When will that occur again?

24) Two runners train at a local high school with an enormous circular track. One of the runners can run once around the circular track in 4 minutes. The other runner can run once around in 6 minutes. If they both begin at the same moment, when will they be together again?

TWO INTERESTING WAYS TO GET L.C.M.s

There is an interesting way to get the least common multiple of two numbers that uses the G.C.D. of the numbers.

Let's try getting the L.C.M. of 8 and 10.

In the method just learned, we write the prime factors of each number.

$$8 = 2 \times 2 \times 2$$

$$10 = 2 \times 5$$

We learned to find both the G.C.D. and the L.C.M. of the two numbers.

The G.C.D. is 2. (since both numbers have one factor of 2 in common)

By our previous work we can see that the L.C.M. is $40 = 2 \times 2 \times 2 \times 5$.

Notice that if we multiply the two numbers, 8×10 and then divide by the G.C.D. 2, we also get 40.

$$\frac{8 \times 10}{2} = 40$$

The shortcut is to look at the product of the two given numbers, and DIVIDE their product by the G.C.D. The result will be the L.C.M.

$$\text{L.C.M.} = \frac{\text{PRODUCT OF THE NUMBERS}}{\text{G.C.D.}}$$

Example 1: Find the L.C.M. of 18 and 45 by this new method.

The G.C.D. of 18 and 45 is 9 since $18 = 2 \times 3^2$ and $45 = 3^2 \times 5$ and they have 3^2 in common.

$$\text{L.C.M.} = \frac{18 \times 45}{9} = 2 \times 45 = 90$$

Example 2: Find the L.C.M. of 8, 12 and 6

To do this problem, we must do only two numbers at a time. First we get the L.C.M. of 8 and 12 (NOTE: the G.C.D. is 4)

$$\text{L.C.M.} = \frac{8 \times 12}{4} = \frac{96}{4} = 24$$

Now we get the L.C.M. of 24 and the remaining number, 6.

$$\text{L.C.M.} = \frac{24 \times 6}{6} = 24$$

The final answer is 24.

It is usually easier to find the LCM of TWO numbers whose G.C.D. is obvious.

Example 3: Find the L.C.M. of 75 and 40.

The G.C.D. of 75 and 40 is 5.

The L.C.M. is $\frac{75 \times 40}{5} = 75 \times \frac{40}{5} = 75 \times 8 = 600$

Another method which you may find easier involves dividing a prime into both numbers until you no longer can. The product of the resulting divisors and the quotients gives the L.C.M. We can call this method the "*division method.*"

Example 4: Find the L.C.M. of 24 and 36.

2) 24, 36	We divide BOTH numbers by a prime
2) 12, 18	until the prime doesn't work.
3) 6, 9	We continue dividing by a prime.
2, 3	

THE PRODUCT OF THE DIVISORS AND THE QUOTIENTS AT THE FINISH GIVES THE L.C.M.

In this problem it is $\underbrace{2 \times 2 \times 3}_{\text{divisors}} \times \underbrace{2 \times 3}_{\text{quotients}} = 72.$

Example 5: Find the L.C.M. of 18 and 45.

3) 18, 45	We divide BOTH numbers by the prime 3.
3) 6, 15	We continue dividing by 3.
2, 5	We can't divide both numbers anymore.

We are left with the divisors 3 and 3 and the quotients 2 and 5.

The L.C.M. is $\underbrace{3 \times 3}_{\text{divisors}} \times \underbrace{2 \times 5}_{\text{quotients}} = 90.$

Example 6: Find the L.C.M. of 8, 12 and 6.

Now we will start with a prime number that divides evenly into at least TWO of the three numbers.

```
2) 8, 12, 6
2) 4,  6, 3
3) 2,  3, 3
   2,  1, 1
```

If a prime doesn't divide evenly into one of the numbers, you bring the number down. (NOTE: we first brought the 3 down. Then we brought the 2 down.)

The L.C.M. is the product of the divisors 2, 2 and 3 and the quotients carried down, 2, 1 and 1.

The L.C.M. is $2 \times 2 \times 3 \times 2 \times 1 \times 1 = 24$.

Example 7: Find the L.C.M. of 12, 20 and 30.

```
2) 12, 20, 30
2)  6, 10, 15
3)  3,  5, 15
5)  1,  5,  5
    1,  1,  1
```

The L.C.M. is $2 \times 2 \times 3 \times 5 = 60$.

TWO INTERESTING WAYS TO GET LCMs EXERCISES:

Practice Set A

Use the G.C.D. method to get the L.C.M. of the following sets of numbers:

1) 12 and 18
2) 10, 6 and 5
3) 27 and 21
4) 18 and 27
5) 42 and 48
6) 18 and 30
7) 100 and 1000
8) 9 and 16
9) 15 and 50
10) 20 and 16
11) 60 and 84
12) 25 and 40

Use the "division method" to find the L.C.M. of the following sets of numbers:

13) 2, 5 and 7
14) 4, 6 and 5
15) 12, 15 and 20
16) 8, 12 and 9
17) 12 and 16
18) 75 and 90
19) 213 and 639
20) 32 and 18
21) 65 and 91
22) 11, 12 and 13
23) 10, 100 and 1000
24) 28, 21, 42 and 6

TWO INTERESTING WAYS TO GET LCMs EXERCISES:

Practice Set B

Use the G.C.D. method to get the L.C.M. of the following sets of numbers:

1) 18 and 21
2) 12 and 30
3) 9 and 15
4) 12 and 10
5) 8 and 14
6) 24 and 18
7) 25 and 60
8) 15, 25 and 45
9) 24 and 32
10) 36 and 216
11) 10, 12 and 15
12) 40 and 48

Use the "division method" to find the L.C.M. of the following sets of numbers:

13) 8, 14 and 21
14) 15, 25 and 30
15) 7, 11 and 13
16) 12, 20 and 16
17) 100 and 40
18) 256 and 192
19) 49 and 343
20) 154 and 110
22) 12, 75 and 10
23) 18, 24, 32 and 15
24) 286 and 182

"SON....TRY TO REMEMBER THAT 'SLASHING THE FURNITURE BY ONE-THIRD' REFERS TO THE SALE PRICE."

CHAPTER 2: FRACTIONS

A CONCRETE APPROACH: INTRODUCTION

Fractions are used to compare parts to a whole. When we take some basic unit (whether it be the traditional grade school pies, or rectangles, or ruler lengths or units of money, etc.) and divide it into small equal *sub*units, we use fractions to describe these subunits.

Consider the following interval to be a UNIT.

If we divide this unit into two equal parts, it looks like this:

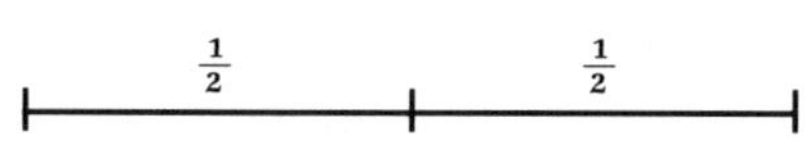

Each part is called a "half", and is denoted by the fraction $\frac{1}{2}$.

We can think of the $\frac{1}{2}$ as meaning "1 out of 2 parts".

The fraction $\frac{2}{2}$ would mean "2 out of 2 parts" which is really the *whole*. Therefore $\frac{2}{2} = 1$.

If we take the same interval and divide it into three equal parts, each part is called a "third" and is denoted by the fraction $\frac{1}{3}$.

The interval would be divided as this:

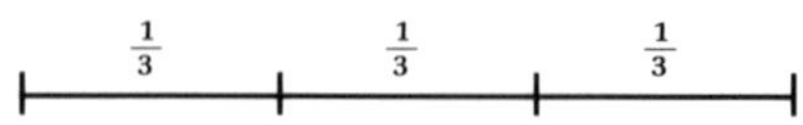

The fraction $\frac{2}{3}$ would mean "2 out of 3 parts". It would be represented by the shaded sections of the following diagram:

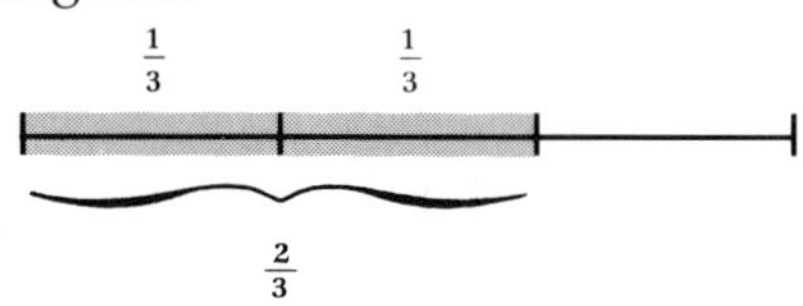

If we take the same unit interval and divide it into four equal parts, each part is called a "fourth", and is denoted by the fraction $\frac{1}{4}$. The whole interval would look like this:

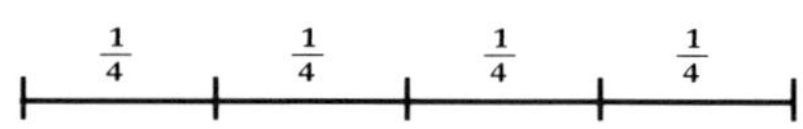

The fraction $\frac{3}{4}$ would mean "3 out of 4 parts", and would appear as 3 of the 4 subintervals being shaded.

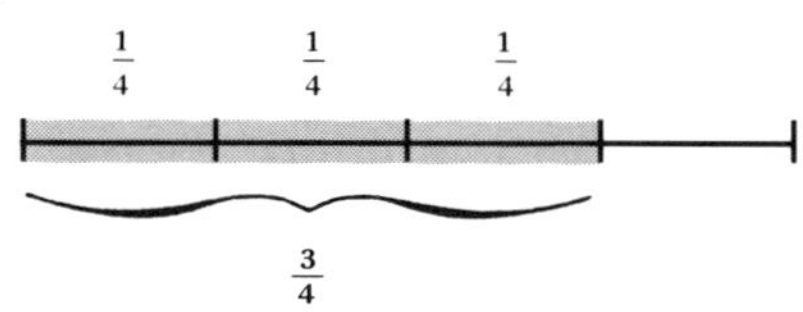

The top part of a fraction is called the *numerator*. It tells how many parts are shaded.

The bottom of a fraction is called the *denominator*. It tells into how many equal parts the unit interval is divided.

Fractions where the numerator is SMALLER than the denominator are called PROPER FRACTIONS. These were the kind most of us were introduced to. We will call the fractions where both numerator and denominator are equal the UNIT FRACTIONS. All unit fractions equal 1 (the whole). For example: $\frac{5}{5} = 1$.

Sometimes we want to describe MORE than a whole. Maybe we want to describe a WHOLE and PART of another whole.

The fraction $\frac{7}{4}$ would mean "7 out of 4 parts" which sounds awkward.

It really means that we divide a whole into 4 equal parts, and then divide another whole into 4 equal parts. Then we consider 7 of the 8 subunits.

Pictorially $\frac{7}{4}$ looks like this:

Fractions where the top (numerator) is LARGER than the bottom (denominator) are called IMPROPER FRACTIONS.

You can see that when dealing with improper fractions, we must be aware of how many *wholes* are implicitly indicated. It is always at least one entire unit (or more) and some subunits. We can determine this very easily if we think of the fraction notation as giving us a division problem.

Example 1: $\frac{7}{4}$ can be thought of as the division problem $4\overline{)7}$ or $7 \div 4$.

If we do this problem, 4 divides into 7 with a quotient of 1 and a remainder of 3.

Literally that means that $\frac{7}{4}$ is *1* whole and *3* parts out of 4. We can write that as $1 + \frac{3}{4}$ or $1\frac{3}{4}$. This form, $1\frac{3}{4}$ is called MIXED NUMBER FORM.

ALL IMPROPER FRACTIONS CAN BE EXPRESSED AS MIXED NUMBERS, AND VICE VERSA.

> To go from IMPROPER FRACTION to MIXED NUMBER:
>
> Divide the denominator *into* the numerator, and write the remainder as a fraction.

Example 2: Change $\frac{11}{3}$ to a mixed number:

Dividing the denominator 3 into the numerator 11, we get 3 with a remainder of 2.

$$\begin{array}{r} 3\,r2 \\ 3\overline{)11} \\ -9 \\ \hline 2 \end{array} \quad \text{or } 3\frac{2}{3}$$

Therefore, $\frac{11}{3} = 3\frac{2}{3}$.

Example 3: Change $\frac{19}{6}$ to a mixed number:

Dividing the denominator 6 into the numerator 19, we get 3 with a remainder of 1.

$$\begin{array}{r} 3\ r1 \\ 6\overline{)19} \\ -\underline{18} \\ 1 \end{array} \qquad \text{or } 3\frac{1}{6}$$

Therefore, $\frac{19}{6} = 3\frac{1}{6}$

> To go from MIXED NUMBER to IMPROPER FRACTION:
>
> Multiply the whole number by the denominator, then add the numerator. That result is written over the denominator.

Example 4: Change $2\frac{5}{8}$ to an improper fraction.

We multiply the whole number 2 by the denominator 8 and add the numerator 5. $(2 \times 8 + 5 = 21)$

The result is $\frac{21}{8}$.

Example 5: Change $9\frac{3}{12}$ to an improper fraction.

$9 \times 12 + 3 = 108 + 3 = 111$, the new numerator.

The result is $\frac{111}{12}$.

FRACTIONS: A CONCRETE APPROACH EXERCISES:

Practice Set A

Change each of the following to MIXED NUMBER form:

1) $\frac{13}{4}$ 2) $\frac{25}{7}$ 3) $\frac{18}{5}$ 4) $\frac{19}{6}$ 5) $\frac{27}{2}$

6) $\frac{18}{7}$ 7) $\frac{100}{9}$ 8) $\frac{11}{5}$ 9) $\frac{1001}{2}$ 10) $\frac{45}{7}$

11) $\frac{89}{10}$ 12) $\frac{35}{12}$ 13) $\frac{36}{5}$ 14) $\frac{1275}{26}$ 15) $\frac{185}{20}$

16) $\frac{42}{3}$ 17) $\frac{17}{15}$ 18) $\frac{98}{11}$ 19) $\frac{41}{3}$ 20) $\frac{200}{3}$

21) $\frac{175}{126}$ 22) $\frac{53}{10}$ 23) $\frac{97}{8}$ 24) $\frac{100}{3}$ 25) $\frac{75}{2}$

Change each of the following to IMPROPER FRACTION form:

26) $10\frac{3}{4}$ 27) $12\frac{9}{10}$ 28) $8\frac{2}{5}$ 29) $1\frac{3}{4}$ 30) $2\frac{1}{2}$

31) $15\frac{3}{4}$ 32) $3\frac{2}{3}$ 33) $11\frac{1}{2}$ 34) $12\frac{1}{8}$ 35) $7\frac{7}{8}$

36) $9\frac{2}{3}$ 37) $29\frac{2}{3}$ 38) $33\frac{1}{3}$ 39) $66\frac{2}{3}$ 40) $12\frac{1}{2}$

41) $5\frac{1}{6}$ 42) $46\frac{7}{8}$ 43) $51\frac{7}{8}$ 44) $1\frac{15}{16}$ 45) $99\frac{9}{10}$

46) $2\frac{3}{4}$ 47) $8\frac{5}{6}$ 48) $101\frac{11}{12}$ 49) $1\frac{1}{2}$ 50) $19\frac{18}{19}$

Shade each figure to represent the given fraction:

51) $\frac{1}{2}$

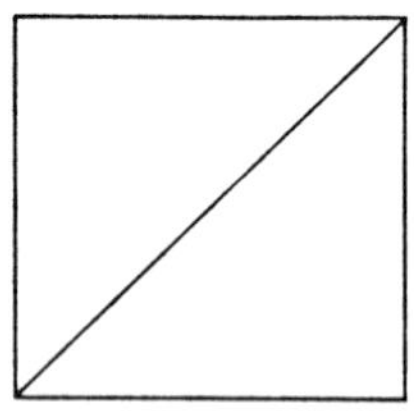

52) $\frac{4}{6}$

53) $\frac{3}{6}$

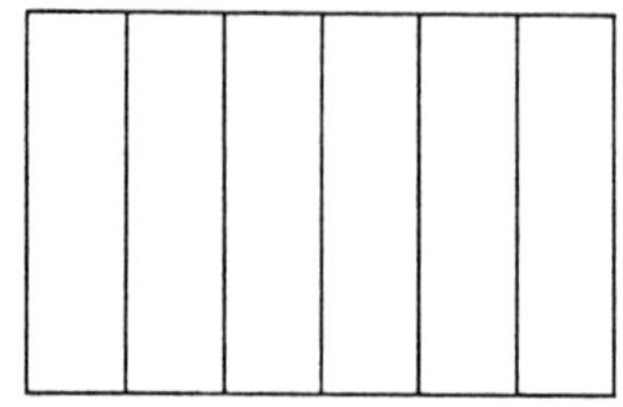

54) $\frac{2}{3}$

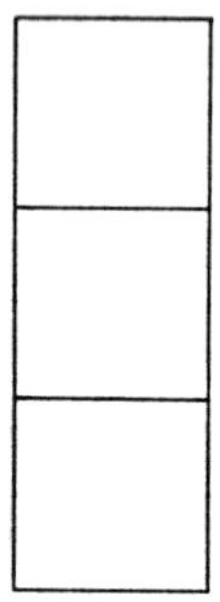

55) $\frac{1}{2}$

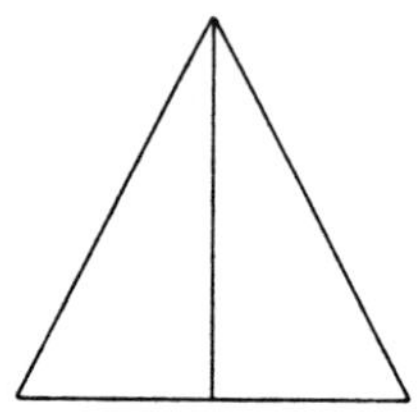

56) $\frac{3}{4}$

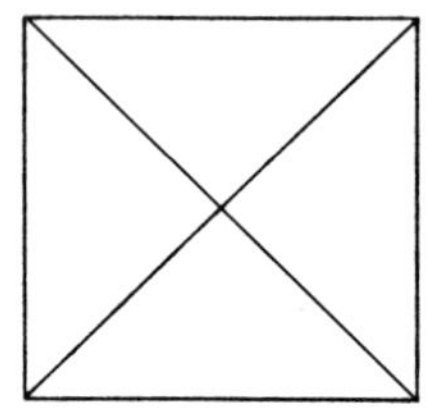

57) Shade $\frac{3}{5}$ of the total region encompassed by these 5 hexagons:

58) Mark one-tenth of the total area in these 5 circles:

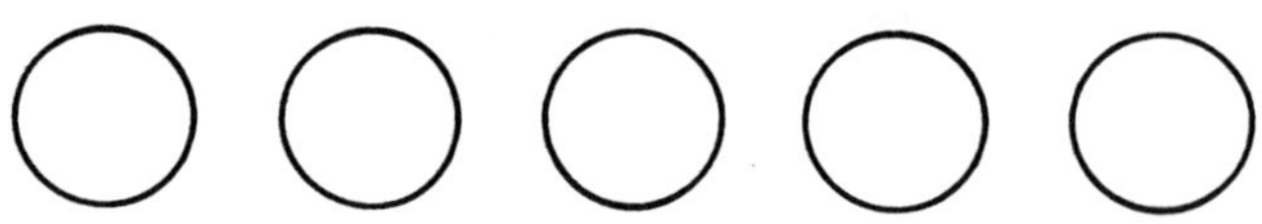

59) Shade $\frac{2}{3}$ of the total area encompassed by these 6 squares:

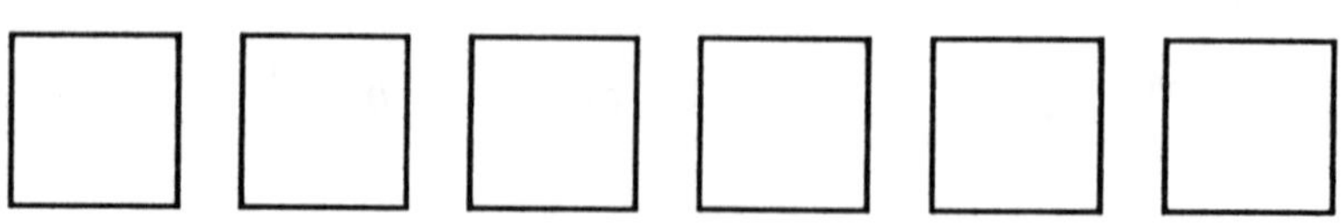

In each of the following pictures, a fraction can be associated with it. Indicate which fraction.

60) 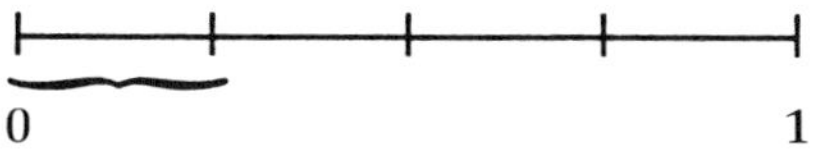

61) 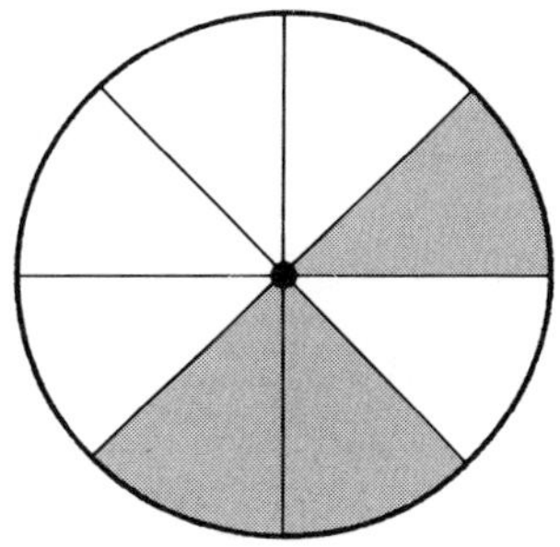

62) 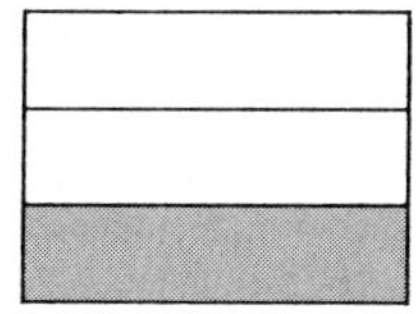

63) 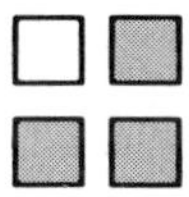

64) 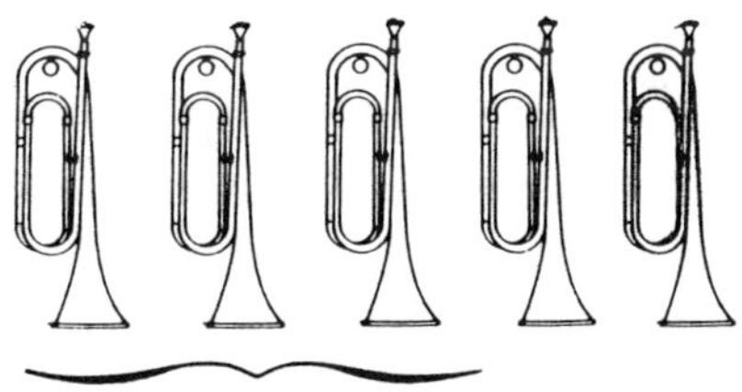

65) 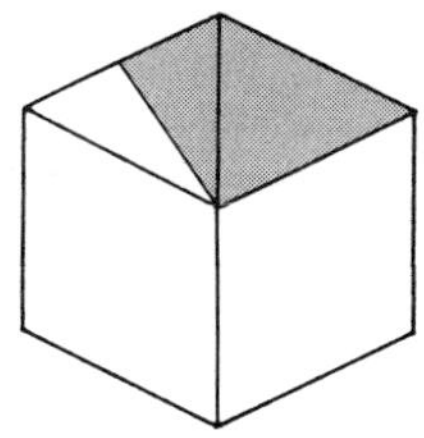

*66) 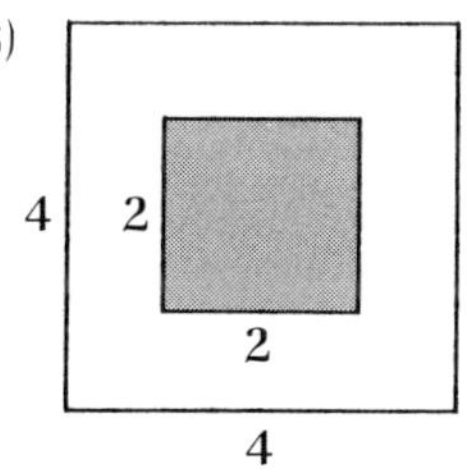

67) 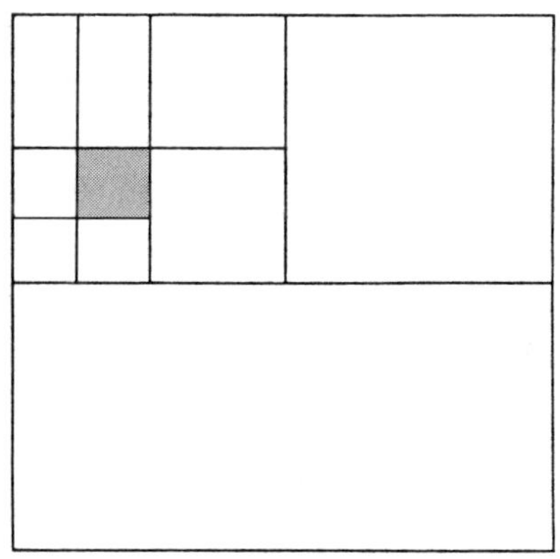

68) 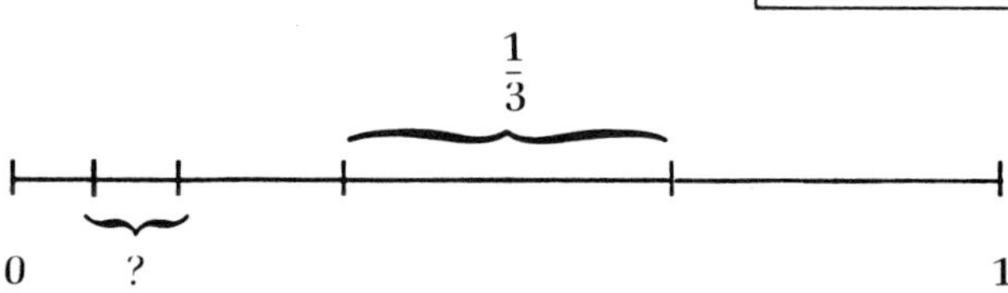

69) 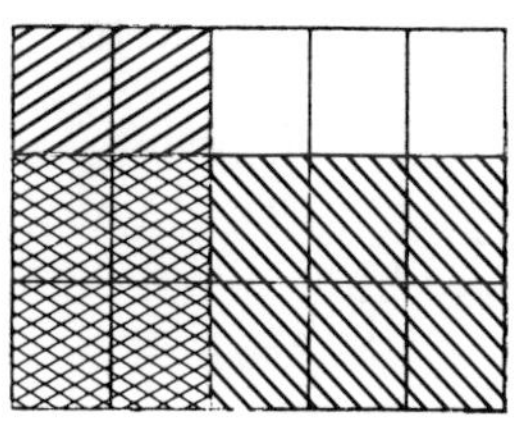

What part of the rectangle is shaded by ////?
What part of rectangle has \\\\ lines?
What part of rectangle has both kinds of lines?

FRACTIONS: A CONCRETE APPROACH EXERCISES:

Practice Set B

Change each of the following to MIXED NUMBER FORM:

1) $\frac{13}{5}$ 2) $\frac{25}{8}$ 3) $\frac{18}{4}$ 4) $\frac{19}{5}$ 5) $\frac{29}{2}$

6) $\frac{19}{7}$ 7) $\frac{200}{11}$ 8) $\frac{13}{6}$ 9) $\frac{2001}{4}$ 10) $\frac{49}{9}$

11) $\frac{789}{100}$ 12) $\frac{44}{3}$ 13) $\frac{48}{13}$ 14) $\frac{1985}{25}$ 15) $\frac{298}{21}$

16) $\frac{48}{4}$ 17) $\frac{23}{22}$ 18) $\frac{101}{11}$ 19) $\frac{43}{3}$ 20) $\frac{142}{3}$

21) $\frac{51}{2}$ 22) $\frac{89}{4}$ 23) $\frac{473}{10}$ 24) $\frac{155}{141}$ 25) $\frac{25}{2}$

Change each of the following to IMPROPER FRACTION form:

26) $8\frac{3}{4}$ 27) $2\frac{2}{3}$ 28) $4\frac{5}{6}$ 29) $10\frac{9}{10}$ 30) $8\frac{1}{2}$

31) $9\frac{3}{5}$ 32) $8\frac{4}{5}$ 33) $6\frac{7}{8}$ 34) $15\frac{1}{3}$ 35) $125\frac{1}{4}$

36) $40\frac{2}{3}$ 37) $786\frac{7}{10}$ 38) $11\frac{1}{11}$ 39) $2\frac{1}{22}$ 40) $100\frac{3}{100}$

41) $9\frac{3}{8}$ 42) $7\frac{7}{8}$ 43) $1\frac{15}{16}$ 44) $9\frac{1}{10}$ 45) $18\frac{7}{8}$

Equivalent Fractions

Two fractions are *equivalent* if they have the same value although their forms are different. For example, the fraction $\frac{1}{2}$ and $\frac{2}{4}$ will denote the same shaded area of the unit interval:

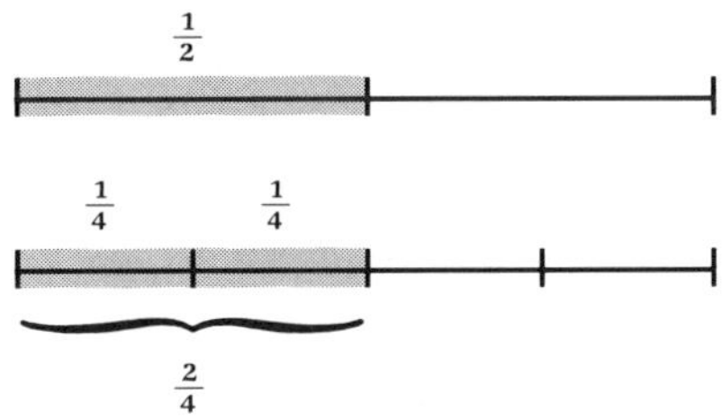

$$\frac{1}{4} + \frac{1}{4} = \frac{2}{4} = \frac{1}{2}$$

One-half of an interval is the same as "2 parts out of 4" of an interval, since 2 is one-half of 4.

By the same reasoning, $\frac{1}{2}$ is the same as $\frac{3}{6}$ since 3 is half of 6.

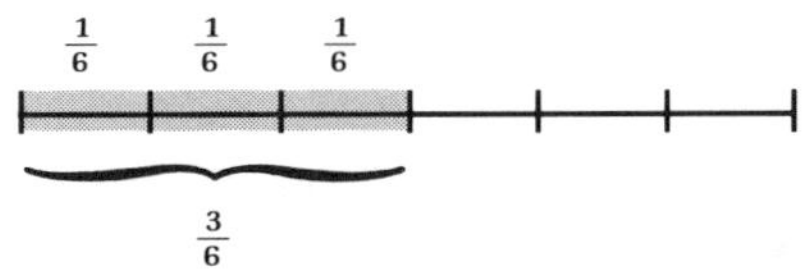

$$\frac{1}{6} + \frac{1}{6} + \frac{1}{6} = \frac{3}{6} = \frac{1}{2}$$

We can clearly see how fractions equivalent to $\frac{1}{2}$ are formed if we think of a rectangle that has been divided into two equal parts, one part shaded. If we divide each part into halves again, we have doubled the number of parts. Now there are four equal parts in total. We must shade two of the four parts to have $\frac{1}{2}$ of the rectangle shaded.

Essentially, we just doubled BOTH the number of parts in total (denominator) and the number of parts shaded (numerator).

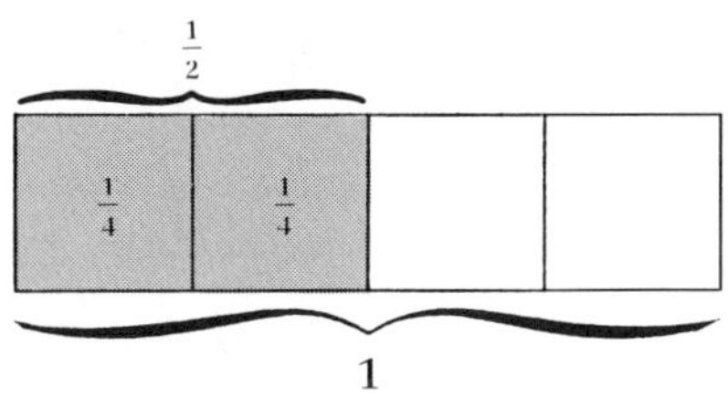

GIVEN ANY FRACTION, WE CAN GENERATE EQUIVALENT FRACTIONS EQUAL TO IT IF WE MULTIPLY *BOTH* NUMERATOR AND DENOMINATOR BY THE SAME NUMBER. WHEN WE DO THIS, IT IS CALLED "**BUILDING FRACTIONS**".

Example 1: Build several fractions equivalent to $\frac{2}{3}$.

$$\frac{2}{3} \xrightarrow{\times 2} \frac{4}{6} \qquad \frac{2}{3} \xrightarrow{\times 5} \frac{10}{15} \qquad \frac{2}{3} \xrightarrow{\times 3} \frac{6}{9} \xrightarrow{\times 2} \frac{12}{18}$$

(×2 numerator and denominator: $\frac{2}{3} = \frac{4}{6}$; ×5: $\frac{2}{3} = \frac{10}{15}$; ×3 then ×2: $\frac{2}{3} \times \frac{6}{9} = \frac{12}{18}$)

Since we are multiplying the top and bottom by the same number, we are really multiplying the fraction by an equivalent of 1.

$$\frac{2}{3} \times \left(\frac{2}{2}\right) = \frac{4}{6} \qquad \frac{2}{3} \times \left(\frac{5}{5}\right) = \frac{10}{15} \qquad \frac{2}{3} \times \left(\frac{6}{6}\right) = \frac{12}{18}$$

$$\left(\frac{2}{2}\right) \to 1 \qquad \left(\frac{5}{5}\right) \to 1 \qquad \left(\frac{6}{6}\right) \to 1$$

Here we are just multiplying by 1.

Example 2: Build several fractions equivalent to $\frac{7}{5}$.

$$\frac{7}{5} = \frac{14}{10} = \frac{35}{25} = \frac{70}{50}$$ etc. at infinitum

(numerator and denominator each $\times\ 2$, $\times\ 5$, $\times\ 10$)

WE CAN ALSO GET FRACTIONS EQUIVALENT TO ANY FRACTION IF WE *DIVIDE* BOTH NUMERATOR AND DENOMINATOR BY THE SAME NUMBER. WHEN WE DO THIS IT IS CALLED "**REDUCING FRACTIONS**."

WHEN WE DIVIDE BOTH THE NUMERATOR AND THE DENOMINATOR BY THEIR GREATEST COMMON DIVISOR (G.C.D.) IT IS CALLED "**REDUCING FRACTIONS TO *LOWEST TERMS***."

Example 3: Reduce $\frac{18}{27}$

We can divide both the 18 and the 27 by the number 3.

Therefore $\frac{18}{27} = \frac{6}{9}$ ($\div\ 3$ top and bottom) (this fraction is reduced, but NOT to lowest terms)

Example 4: Reduce $\frac{18}{27}$ to *lowest terms*.

The G.C.D. of 18 and 27 is 9 ($18 = 2 \times 3^2$ and $27 = 3^3$)

We can divide both the 18 and the 27 by 9 to get the most reduced form of the equivalent fraction.

$\frac{18}{27} = \frac{2}{3}$ ($\div\ 9$ top and bottom) $\frac{2}{3}$ is the *most reduced form*.

Example 5: Reduce $\frac{51}{81}$ to lowest terms.

It is not readily obvious what the G.C.D. of 51 and 81 is. We must look at the prime factors of both to determine this.

$51 = 3 \times 17$ and $81 = 3 \times 3 \times 3 \times 3$. The greatest common divisor is 3.

We divide both the 51 and the 81 by the 3.

$$\frac{51}{81} = \frac{17}{27} \quad (\div 3 \text{ top and bottom})$$

Therefore $\frac{17}{27}$ is the equivalent fraction reduced to *lowest* terms.

Example 6: Reduce $\frac{48}{60}$ to lowest terms.

It may not be obvious what the G.C.D. of 48 and 60 is. We must look at the prime factors of both numbers to determine this.

$$48 = 2 \times 2 \times 2 \times 2 \times 3 \quad \text{and} \quad 60 = 2 \times 2 \times 3 \times 5$$

The greatest common divisor is $2 \times 2 \times 3$ or 12. We divide both the 48 and the 60 by 12.

$$\frac{48}{60} = \frac{4}{5} \quad (\div 12 \text{ top and bottom})$$

Therefore, $\frac{4}{5}$ is the equivalent fraction reduced to *lowest* terms.

Example 7: Reduce $\frac{36}{55}$ to lowest terms.

We must find the G.C.D. of 36 and 55.

$36 = 2 \times 2 \times 3 \times 3 \qquad 55 = 5 \times 11$

The greatest common divisor is 1. The numbers are relatively prime.

Therefore, the number $\frac{36}{55}$ is *already reduced*.

There is a simple rule to tell whether two fractions are equivalent. We simply cross multiply. The numerator of one fraction times the denominator of the other *should equal* the numerator of the other fraction times the denominator of the one.

Example 8: Show that $\frac{3}{4} = \frac{12}{16}$

$\frac{3}{4} = \frac{12}{16}$ $\quad 3 \times 16 = 4 \times 12 = 48$

EQUIVALENT FRACTIONS EXERCISES:

Practice Set A

For each of the following, build 5 equivalent fractions:

1) $\frac{1}{3}$ 2) $\frac{2}{5}$ 3) $\frac{5}{3}$ 4) $\frac{3}{2}$ 5) $\frac{3}{4}$

6) $\frac{7}{8}$ 7) $\frac{1}{2}$ 8) $\frac{3}{10}$ 9) $\frac{1}{8}$ 10) $\frac{1}{6}$

11) $\frac{5}{6}$ 12) $\frac{8}{15}$ 13) $\frac{9}{10}$ 14) $\frac{2}{3}$ 15) $\frac{7}{9}$

Write these following numbers in terms of the new numerators and denominators as indicated:

16) $\frac{3}{4} = \frac{?}{8}$ 17) $\frac{4}{5} = \frac{?}{25}$ 18) $\frac{5}{6} = \frac{?}{24}$

19) $\frac{1}{2} = \frac{?}{98}$ 20) $\frac{3}{5} = \frac{51}{?}$ 21) $\frac{7}{9} = \frac{77}{?}$

22) $\frac{0}{8} = \frac{?}{16}$ 23) $\frac{0}{8} = \frac{?}{9}$ 24) $\frac{7}{5} = \frac{?}{55}$

25) $\frac{2}{3} = \frac{?}{39}$ 26) $\frac{5}{8} = \frac{?}{1000}$ 27) $\frac{2}{5} = \frac{24}{?}$

28) $\frac{8}{5} = \frac{24}{?}$ 29) $1\frac{1}{2} = \frac{?}{2} = \frac{?}{4}$ 30) $3\frac{3}{5} = \frac{?}{5} = \frac{?}{10}$

31) $15\frac{1}{3} = \frac{?}{3} = \frac{?}{12}$ 32) $8\frac{4}{5} = \frac{?}{5} = \frac{?}{10}$ 33) $2\frac{2}{3} = \frac{?}{3} = \frac{?}{15}$

34) $6\frac{7}{8} = \frac{?}{8} = \frac{?}{32}$ 35) $4\frac{2}{3} = \frac{?}{9}$

Reduce each of the following fractions to lowest terms:

36) $\frac{28}{35}$ 37) $\frac{32}{48}$ 38) $\frac{22}{77}$ 39) $\frac{18}{36}$ 40) $\frac{12}{32}$

41) $\frac{6}{9}$ 42) $\frac{16}{18}$ 43) $\frac{9}{36}$ 44) $\frac{90}{110}$ 45) $\frac{75}{100}$

46) $\frac{42}{72}$ 47) $\frac{9}{15}$ 48) $\frac{10}{4}$ 49) $\frac{24}{18}$ 50) $\frac{144}{18}$

51) $\frac{35}{60}$ 52) $\frac{55}{80}$ 53) $\frac{76}{15}$ 54) $\frac{185}{20}$ 55) $\frac{9}{75}$

56) $\frac{75}{175}$ 57) $\frac{16}{81}$ 58) $\frac{0}{11}$ 59) $\frac{21}{30}$ 60) $\frac{51}{72}$

Mark each of the following as TRUE or FALSE. If false, correct the numerator of the fraction on the right to make the fractions equivalent:

61) $\frac{2}{5} = \frac{24}{60}$ 62) $\frac{7}{8} = \frac{56}{64}$ 63) $\frac{5}{8} = \frac{15}{32}$

64) $\frac{12}{35} = \frac{60}{70}$ 65) $\frac{1}{3} = \frac{33}{66}$ 66) $\frac{9}{20} = \frac{45}{100}$

67) $\frac{5}{4} = \frac{75}{100}$ 68) $\frac{3}{2} = \frac{51}{36}$ 69) $\frac{24}{5} = \frac{76}{15}$

70) $\frac{5}{6} = \frac{10}{12} = \frac{12}{18} = \frac{40}{48}$

(correct this)

EQUIVALENT FRACTIONS EXERCISES

Practice Set B

For each of the following build 5 equivalent fractions:

1) $\frac{1}{4}$ 2) $\frac{3}{5}$ 3) $\frac{7}{3}$ 4) $\frac{4}{3}$ 5) $\frac{6}{5}$

6) $\frac{5}{8}$ 7) $\frac{1}{5}$ 8) $\frac{7}{10}$ 9) $\frac{1}{9}$ 10) $\frac{9}{11}$

11) $\frac{1}{8}$ 12) $\frac{9}{16}$ 13) $\frac{5}{2}$ 14) $\frac{3}{13}$ 15) $\frac{7}{12}$

Write the following numbers in terms of the new numerators and denominators, as indicated:

16) $\frac{3}{4} = \frac{?}{28}$ 17) $\frac{4}{5} = \frac{16}{?}$ 18) $\frac{5}{6} = \frac{?}{66}$

19) $\frac{1}{2} = \frac{224}{?}$ 20) $\frac{2}{5} = \frac{38}{?}$ 21) $\frac{9}{7} = \frac{?}{35}$

22) $\frac{0}{9} = \frac{?}{27}$ 23) $\frac{0}{9} = \frac{?}{10}$ 24) $\frac{7}{8} = \frac{28}{?}$

25) $\frac{2}{3} = \frac{?}{51}$ 26) $\frac{9}{25} = \frac{?}{1000}$ 27) $\frac{3}{5} = \frac{42}{?}$

28) $\frac{6}{5} = \frac{42}{?}$ 29) $2\frac{1}{2} = \frac{?}{2} = \frac{?}{6}$ 30) $3\frac{5}{8} = \frac{?}{8} = \frac{?}{24}$

31) $4\frac{4}{5} = \frac{?}{5} = \frac{?}{25}$ 32) $33\frac{1}{3} = \frac{?}{3} = \frac{?}{12}$ 33) $12\frac{2}{5} = \frac{?}{5} = \frac{?}{10}$

34) $10\frac{1}{8} = \frac{?}{40}$ 35) $66\frac{2}{3} = \frac{?}{9}$

Reduce each of the following fractions to lowest terms:

36) $\frac{8}{10}$ 37) $\frac{24}{32}$ 38) $\frac{11}{121}$ 39) $\frac{8}{12}$ 40) $\frac{12}{15}$

41) $\frac{9}{6}$ 42) $\frac{15}{16}$ 43) $\frac{30}{100}$ 44) $\frac{6}{18}$ 45) $\frac{150}{200}$

46) $\frac{9}{9}$ 47) $\frac{6}{10}$ 48) $\frac{33}{99}$ 49) $\frac{15}{12}$ 50) $\frac{115}{100}$

51) $\frac{14}{21}$ 52) $\frac{256}{112}$ 53) $\frac{15}{25}$ 54) $\frac{48}{88}$ 55) $\frac{108}{144}$

Mark each of the following as TRUE or FALSE. If false, correct the numerator of the fraction on the right to make the fractions equivalent:

56) $\frac{8}{16} = \frac{3}{4}$ 57) $\frac{25}{32} = \frac{75}{92}$ 58) $\frac{5}{4} = \frac{15}{12}$

59) $\frac{6}{10} = \frac{48}{70}$ 60) $\frac{2}{3} = \frac{35}{51}$

GRAPHING FRACTIONS ON A NUMBER LINE AND COMPARING THE SIZE OF FRACTIONS

We can get a good idea of the relative size of any number if we draw a number line to indicate all the positive real numbers.

We draw a line, and start the first number at 0. We then proceed to divide the line into intervals of EQUAL lengths, and label each division with the consecutive counting numbers.

We can place a fraction on the number line once we determine between what numbers it lies. We divide that interval into the appropriate number of equal subintervals (depending on the denominator). We then label the subintervals consecutively.

Example 1: The fraction $\frac{1}{2}$ is between the whole numbers 0 and 1.

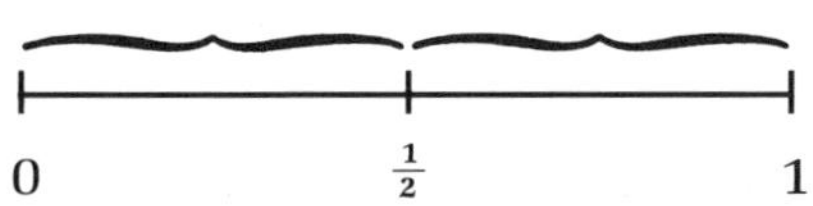

The denominator, 2, tells us to divide the interval between 0 and 1 into TWO equal parts.

Example 2: Graph the fraction $\frac{3}{4}$ on the number line.

The fraction $\frac{3}{4}$ is between 0 and 1. The denominator, 4, tells us to divide the interval into FOUR equal parts.

We then label each part consecutively as $\frac{1}{4}, \frac{2}{4}, \frac{3}{4}$, and $\frac{4}{4} = 1$.

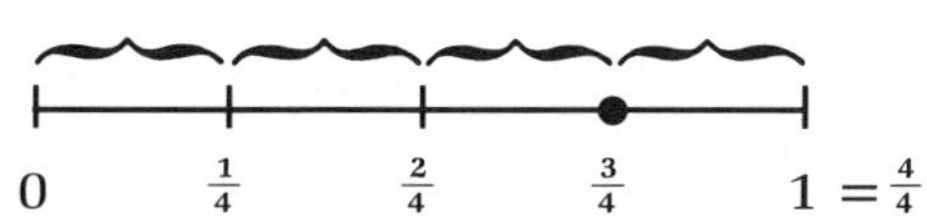

$\frac{3}{4}$ is the *third* division in the interval.

Example 3: Graph the fraction $\frac{7}{3}$ on the number line:

The fraction $\frac{7}{3}$ is in between the whole numbers 2 and 3. The fraction $\frac{7}{3}$ is really the mixed number $2\frac{1}{3}$.

We divide the interval between 2 and 3 into THREE equal parts, since the denominator is 3.

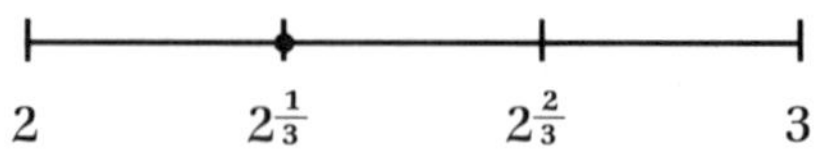

Since $\frac{7}{3} = 2\frac{1}{3}$, it is the *first* division in the interval.

Example 4: Graph the fraction $\frac{12}{5}$ on the number line.

The fraction $\frac{12}{5}$ is also between the whole numbers 2 and 3, since it is the mixed number $2\frac{2}{5}$.

The denominator, 5, tells us to divide the interval into FIVE equal parts. $\frac{12}{5}$ is the *second* division in the interval since $\frac{12}{5} = 2\frac{2}{5}$.

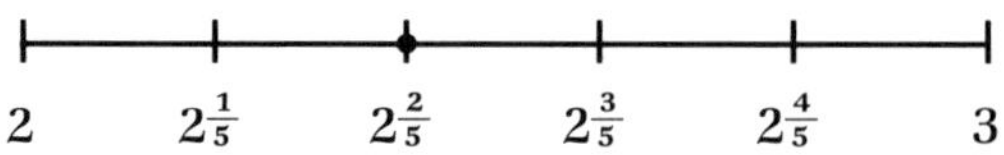

Sometimes it is necessary to compare the size of several fractions. For example, $\frac{7}{3}$ and $\frac{12}{5}$ are both in the interval between the whole numbers 2 and 3. Both fractions are closer to 2 than to 3. How can we tell which one is larger?

If we use a ruler and accurately estimate where the thirds are and where the fifths are in the interval, we would see that $\frac{12}{5}$ is slightly larger than $\frac{7}{3}$.

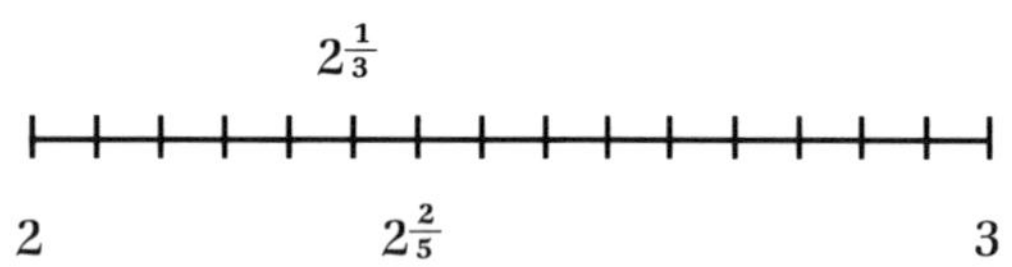

$\frac{12}{5} = 2\frac{2}{5}$ is larger than $\frac{7}{3} = 2\frac{1}{3}$

Measuring out the fractions using a ruler is too time consuming or it is not accurate enough because of the limitation of the ruler used. We need a more efficient way to determine these comparisons.

COMPARE THE SIZE OF SEVERAL FRACTIONS BY CHANGING THEM TO EQUIVALENT FRACTIONS, ALL OF WHOSE DENOMINATORS ARE THE L.C.M. OF THE GIVEN DENOMINATORS.

Example 5: Which is larger, $\frac{7}{3}$ or $\frac{12}{5}$?

The L.C.M. of the denominators 3 and 5 is 15.

$$\frac{7}{3} = \frac{35}{15} \quad \text{and} \quad \frac{12}{5} = \frac{36}{15}$$

$$\frac{12}{5} \text{ is greater than } \frac{7}{3} \quad \left(\frac{12}{5} > \frac{7}{3}\right)$$

$$\downarrow \qquad \downarrow$$

$$\text{since } \frac{36}{15} \text{ is greater than } \frac{35}{15}$$

Example 6: Which is smallest: $\frac{1}{2}$ $\frac{3}{4}$, or $\frac{5}{6}$?

We first must find the L.C.M. of the denominators 2, 4 and 6.

Looking at their prime factors, we see the L.C.M. is 12.

$$\begin{aligned} 2 &= 2 \\ 4 &= 2 \times 2 \\ 6 &= 2 \times 3 \\ \hline \text{L.C.M.} &= 2 \times 2 \times 3 = 12 \end{aligned}$$

Next we change each fraction to an equivalent one with new denominator of 12.

$$\frac{1}{2} \overset{\times 6}{=} \frac{6}{12} \qquad \frac{3}{4} \overset{\times 3}{=} \frac{9}{12} \qquad \frac{5}{6} \overset{\times 2}{=} \frac{10}{12}$$

We can see by comparing the equivalent fractions that

$\frac{1}{2} < \frac{3}{4} < \frac{5}{6}$ (read as "$\frac{1}{2}$ is less than $\frac{3}{4}$ which is less than $\frac{5}{6}$")

$\left(\frac{6}{12} < \frac{9}{12} < \frac{10}{12}\right)$ (the symbol $<$ is read as "less than")

The fraction $\frac{1}{2}$ is the smallest.

Example 7: Graph $\frac{1}{2}$, $\frac{3}{4}$ and $\frac{5}{6}$ on the number line.

Since $\frac{1}{2}$, $\frac{3}{4}$ and $\frac{5}{6}$ can all be changed to twelfths, we actually graph $\frac{1}{2} = \frac{6}{12}$, $\frac{3}{4} = \frac{9}{12}$, and $\frac{5}{6} = \frac{10}{12}$.

All the numbers lie between 0 and 1.

We divide that interval into 12 equal parts, and number each division accordingly.

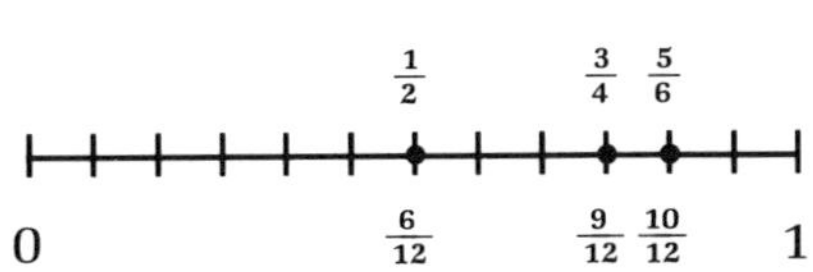

Example 8: Which of the following fractions are between $\frac{1}{2}$ and 1: $\frac{4}{7}$ or $\frac{5}{11}$?

We must see which of the fractions are larger than $\frac{1}{2}$, yet smaller than 1. One way to work this problem is to compare each fraction's size to $\frac{1}{2}$.

Comparing $\frac{4}{7}$ to $\frac{1}{2}$, we must change both fractions to their equivalents with denominator 14.

$$\frac{4}{7} = \frac{8}{14} \ (\times 2) \quad \text{and} \quad \frac{1}{2} = \frac{7}{14} \ (\times 7)$$

Since $\frac{8}{14}$ is larger than $\frac{7}{14}$, then $\frac{4}{7}$ is larger than $\frac{1}{2}$.
So $\frac{4}{7}$ is between $\frac{1}{2}$ and 1.

Comparing $\frac{5}{11}$ to $\frac{1}{2}$, we must change both fractions to their equivalent with denominator 22.

$$\frac{5}{11} = \frac{10}{22} \ (\times 2) \qquad \frac{1}{2} = \frac{11}{22} \ (\times 11)$$

Since $\frac{10}{22}$ is smaller than $\frac{11}{22}$, then $\frac{5}{11}$ is smaller than $\frac{1}{2}$.

So $\frac{5}{11}$ is *NOT* between $\frac{1}{2}$ and 1.

GRAPHING FRACTIONS AND ORDERING FRACTIONS EXERCISES:

Practice Set A

For each of the following groups of fractions, draw the appropriate intervals on the number line and GRAPH the numbers. Do a separate number line for each problem.

1) $\frac{1}{2}, \frac{3}{4}$

2) $\frac{1}{2}, \frac{3}{4}, \frac{7}{8}$

3) $\frac{1}{2}, \frac{3}{4}, \frac{2}{3}$

4) $\frac{2}{3}, \frac{5}{6}$

5) $\frac{2}{3}, \frac{5}{6}, \frac{1}{2}$ and $\frac{11}{12}$

6) $1\frac{1}{2}, \frac{5}{3}, \frac{7}{6}$

7) $\frac{3}{4}, \frac{7}{8}, \frac{5}{3}, \frac{17}{5}, {*}\frac{125}{50}$

Between what two whole numbers do the following fractions lie? To what whole number is each fraction closer?

8) $\frac{18}{5}$

9) $\frac{17}{4}$

10) $\frac{9}{8}$

11) $\frac{7}{9}$

12) $\frac{20}{3}$

13) $\frac{125}{50}$

14) $\frac{99}{12}$

15) $\frac{55}{99}$

Order the following fractions from smallest to largest. List them in that order.

16) $\frac{2}{3}, \frac{7}{8}, \frac{5}{6}$ and $\frac{11}{12}$

17) $\frac{7}{8}, \frac{13}{14}, \frac{2}{3}$ and $\frac{15}{16}$

18) $\frac{9}{10}, \frac{9001}{10000}, \frac{99}{100}, \frac{901}{1000}, \frac{909}{1000}$

19) $\frac{19}{7}, \frac{5}{2}, \frac{28}{11}$

20) $\frac{11}{7}, \frac{13}{9}, \frac{28}{18}, \frac{17}{11}$ and $\frac{3}{2}$

Which of the following fractions are between 0 and $\frac{1}{2}$? (Call the fraction x. $0 < x < \frac{1}{2}$)

21) $\frac{1}{3}$

22) $\frac{2}{5}$

23) $\frac{3}{4}$

24) $\frac{7}{9}$

25) $\frac{4}{9}$

26) $\frac{4}{7}$

27) $\frac{11}{21}$

28) $\frac{19}{39}$

29) $\frac{50}{99}$

30) $\frac{49}{99}$

Which of the following fractions are between $\frac{1}{2}$ and 1? ($\frac{1}{2} < x < 1$)

31) $\frac{2}{3}$

32) $\frac{3}{4}$

33) $\frac{7}{15}$

34) $\frac{11}{19}$

35) $\frac{11}{12}$

36) $\frac{4}{9}$

Which of the following fractions are between 0 and $\frac{1}{3}$? $(0 < x < \frac{1}{3})$

37) $\frac{2}{9}$ 38) $\frac{7}{24}$ 39) $\frac{14}{39}$ 40) $\frac{99}{300}$ 41) $\frac{11}{32}$

42) $\frac{3}{4}$ 43) $\frac{7}{15}$ 44) $\frac{5}{19}$

Which of the following fractions are between $\frac{1}{2}$ and $\frac{3}{4}$? $(\frac{1}{2} < x < \frac{3}{4})$

45) $\frac{2}{3}$ 46) $\frac{7}{8}$ 47) $\frac{9}{16}$ 48) $\frac{8}{15}$ 49) $\frac{3}{5}$

50) $\frac{3}{7}$

51) List five fractions that are greater than 1 yet less than $1\frac{1}{2}$.

52) List five fractions that are greater than $1\frac{1}{2}$ yet smaller than $1\frac{7}{8}$.

53) List five fractions that are greater than $2\frac{1}{3}$ yet less than $2\frac{2}{3}$.

GRAPHING FRACTIONS AND ORDERING FRACTIONS

Practice Set B

For each of the following groups of fractions, draw the appropriate intervals on the number line and GRAPH the numbers. Do a separate number line for each problem.

1) $\frac{1}{2}, \frac{1}{6}, \frac{5}{6}$ 2) $\frac{1}{3}, \frac{5}{6}$ 3) $\frac{1}{4}, \frac{1}{2}, \frac{3}{4}$

4) $2\frac{1}{2}, 2\frac{3}{4}, 2\frac{7}{8}$ 5) $\frac{1}{2}, \frac{2}{3}, \frac{5}{3}, \frac{17}{3}, 5\frac{1}{2}$ 6) $\frac{1}{4}, \frac{5}{4}, \frac{5}{2}, \frac{11}{8}$

7) $1\frac{1}{3}, 1\frac{1}{4}, 1\frac{7}{12}$

Between what two whole numbers do the following fractions lie? To what whole number is each fraction closer?

8) $\frac{17}{3}$ 9) $\frac{18}{15}$ 10) $\frac{99}{100}$ 11) $\frac{12}{8}$ 12) $\frac{100}{33}$

13) $\frac{200}{3}$ 14) $\frac{375}{150}$ 15) $\frac{1000}{99}$

Order the following fractions from smallest to largest. List them in that order.

16) $\frac{3}{2}, \frac{5}{3}, \frac{11}{6}$ 17) $\frac{9}{10}, \frac{4}{5}, \frac{14}{15}$

18) $\frac{3}{10}, \frac{3001}{10000}, \frac{301}{1000}, \frac{31}{100}, \frac{1}{3}$ 19) $\frac{14}{9}, \frac{20}{13}, \frac{8}{5}$

20) $\frac{5}{12}, \frac{1}{2}, \frac{11}{24}, \frac{23}{48}, \frac{4}{9}$

Which of the following fractions are between 0 and $\frac{1}{2}$? (Call the fraction x. $0 < x < \frac{1}{2}$)

21) $\frac{8}{18}$ 22) $\frac{17}{35}$ 23) $\frac{6}{11}$ 24) $\frac{13}{25}$ 25) $\frac{49}{99}$

26) $\frac{7}{15}$ 27) $\frac{7}{12}$ 28) $\frac{435}{1000}$ 29) $\frac{2001}{4000}$ 30) $\frac{9}{16}$

Which of the following fractions are between $\frac{1}{2}$ and 1? ($\frac{1}{2} < x < 1$)

31) $\frac{5}{9}$ 32) $\frac{4}{7}$ 33) $\frac{15}{31}$ 34) $\frac{28}{55}$ 35) $\frac{501}{1000}$

36) $\frac{7}{12}$

Which of the following fractions are between 0 and $\frac{1}{3}$? ($0 < x < \frac{1}{3}$)

37) $\frac{5}{12}$ 38) $\frac{11}{36}$ 39) $\frac{19}{60}$ 40) $\frac{33}{100}$ 41) $\frac{15}{44}$

42) $\frac{1}{4}$ 43) $\frac{2}{5}$ 44) $\frac{10}{33}$

Which of the following fractions are between $\frac{1}{2}$ and $\frac{3}{4}$? ($\frac{1}{2} < x < \frac{3}{4}$)

45) $\frac{4}{5}$ 46) $\frac{5}{8}$ 47) $\frac{7}{10}$ 48) $\frac{5}{9}$ 49) $\frac{7}{12}$

50) $\frac{49}{99}$

51) List five fractions that are greater than 1 yet less than $1\frac{1}{3}$.

52) List five fractions that are greater than $2\frac{1}{2}$ yet less than $2\frac{7}{8}$.

ADDITION AND SUBTRACTION OF FRACTIONS

Fractions can only be added or subtracted if they have the SAME denominator. If a problem is given where the denominators are different, all the fractions have to be changed to equivalent ones with the *same* denominator. This makes sense if we think of adding fractions as just counting LIKE parts.

For example: $\frac{1}{3} + \frac{1}{3} = \frac{2}{3}$

Adding one-*third* and one-*third* gives two-*thirds*.

Similarly, adding three-*fifths* and four-*fifths* will give seven-*fifths*.

$$\frac{3}{5} + \frac{4}{5} = \frac{7}{5}.$$

If all fractions in a problem have the same denominator, then all we are doing is *counting* parts.

Example 1: Add $\frac{1}{9} + \frac{2}{9} + \frac{4}{9}$.

We are adding 1-*ninth* and 2-*ninths* and 4-*ninths*. The answer will be 7-*ninths*. ($\frac{7}{9}$)

We simply added the numerators 1 + 2 + 4, since all the fractions were *ninths*.

Example 2: Compute $\frac{3}{5} + \frac{2}{5} + \frac{4}{5} - \frac{6}{5}$

We simply add/subtract the numerators since the problem involves only fifths.

$$\frac{(3 + 2 + 4) - 6}{5} = \frac{9 - 6}{5} = \frac{3}{5}$$

The answer is $\frac{3}{5}$.

RULE 1:	WHEN ALL THE FRACTIONS IN AN EXERCISE HAVE THE SAME DENOMINATOR, WE SIMPLY ADD OR SUBTRACT THE NUMERATORS. OUR ANSWER WILL HAVE THE SAME DENOMINATOR AS APPEARS IN THE EXERCISE.

Example 3: $\frac{5}{12} + \frac{7}{12} - \frac{2}{12}$

Adding and subtracting the numerators, we get:

$$\frac{5 + 7 - 2}{12} = \frac{10}{12}.$$

This would be the answer except that all fraction results must be reduced to lowest terms. Doing this we get $\frac{5}{6}$ as the answer.

$$\left(\frac{10}{12} \overset{\div 2}{=} \frac{5}{6}\right) \quad (\div 2 \text{ numerator and denominator})$$

Example 4: Add $\frac{4}{5} + \frac{3}{4}$.

Since the denominators are different we must first change both fractions to equivalent fractions with new denominators that are the same. We do this by getting the L.C.M. of the original denominators.

The L.C.M. of 5 and 4 is 20.

$$\frac{4}{5} = \frac{16}{20} \ (\times 4) \quad \text{and} \quad \frac{3}{4} = \frac{15}{20} \ (\times 5)$$

Now we add the twentieths.

$$\frac{4}{5} + \frac{3}{4}$$

$$\downarrow \quad \downarrow$$

$$\frac{16}{20} + \frac{15}{20} = \frac{31}{20} = 1\frac{11}{20}.$$

Our answer is $\frac{31}{20}$ or $1\frac{11}{20}$.

RULE 2: WHEN THE FRACTIONS IN AN EXERCISE HAVE DIFFERENT DENOMINATORS, WE FIRST GET THE L.C.M. OF THE DENOMINATORS. (THIS IS OFTEN CALLED THE LEAST COMMON DENOMINATOR, L.C.D.) SECONDLY, WE CHANGE EACH FRACTION TO ITS EQUIVALENT WITH THE L.C.M. AS DENOMINATOR. THIRDLY, WE ADD OR SUBTRACT THE NUMERATORS OF THE EQUIVALENT FRACTIONS. OUR ANSWER WILL HAVE THE L.C.M. AS THE DENOMINATOR.

Example 5: Compute $\frac{3}{5} + \frac{2}{3} - \frac{3}{4}$

Firstly, we find the L.C.M. of 5, 3 and 4.

The L.C.M. is 60. ($5 = 5$, $3 = 3$, $4 = 2^2$, L.C.M. $= 2^2 \times 3 \times 5 = 60$)

Secondly. we get the equivalent fractions.

$$\frac{3}{5} \xrightarrow{\times 12} = \frac{36}{60}; \quad \frac{2}{3} \xrightarrow{\times 20} = \frac{40}{60}; \quad \frac{3}{4} \xrightarrow{\times 15} = \frac{45}{60}$$

Thirdly, we add and subtract the new numerators.

$$\frac{36}{60} + \frac{40}{60} - \frac{45}{60} = \frac{36 + 40 - 45}{60} = \frac{31}{60}$$

Our answer is $\frac{31}{60}$.

Example 6: $3\frac{3}{8} + 2\frac{1}{2}$

First we get the L.C.M. of the denominators. This least common denominator is 8.

Second we get equivalent fractions.

$$\frac{3}{8} = \frac{3}{8} \quad \frac{1}{2} = \frac{4}{8}$$

At the end we add the numerators of the fraction parts. Then we add the whole numbers in their column.

$$\begin{array}{r} 3\frac{3}{8} \\ + \\ 2\frac{1}{2} \\ \hline \end{array} = \begin{array}{r} 3\frac{3}{8} \\ + \\ 2\frac{4}{8} \\ \hline 5\frac{7}{8} \end{array}$$

The answer is $5\frac{7}{8}$

Example 7: $2\frac{3}{4} + 3\frac{1}{2} + 1\frac{2}{3}$

First we get the L.C.M. of the denominators 2, 3, and 4. (L.C.M. = 12)

Second, we get equivalent fractions.

$$\frac{3}{4} \xrightarrow{\times 3} = \frac{9}{12}; \quad \frac{1}{2} \xrightarrow{\times 6} = \frac{6}{12}; \quad \frac{2}{3} \xrightarrow{\times 4} = \frac{8}{12}$$

Third, we add numerators. Then we add the whole numbers separately.

$$\begin{array}{rcl} 2\frac{3}{4} & = & 2\frac{9}{12} \\ 3\frac{1}{2} & = & 3\frac{6}{12} \\ 1\frac{2}{3} & = & 1\frac{8}{12} \\ \hline & & 6\frac{23}{12} \end{array} = 6 + 1\frac{11}{12} = 6 + 1 + \frac{11}{12} = 7\frac{11}{12}.$$

The answer is $7\frac{11}{12}$

Example 8: $9\frac{7}{12} - 6\frac{5}{12}$

Since the fractions in this exercise have the same denominator, we need only subtract their numerators. Then we deal with the whole numbers.

$$\begin{array}{r} 9\frac{7}{12} \\ -\quad 6\frac{5}{12} \\ \hline 3\frac{2}{12} \end{array}$$

The answer must be reduced. So $3\frac{2}{12} = 3\frac{1}{6}$.

Example 9: $11\frac{5}{6} - 5\frac{3}{4}$

We must first change the fraction to equivalent fractions with the L.C.D. of 12.

$$\frac{5}{6} \xrightarrow{\times 2} \frac{10}{12} \quad \text{and} \quad \frac{3}{4} \xrightarrow{\times 3} \frac{9}{12}$$

Then we proceed to subtract the new numerators and subtract the whole numbers.

$$\begin{array}{r} 11\frac{5}{6} \\ -\quad 5\frac{3}{4} \\ \hline \end{array} \quad \rightarrow \quad \begin{array}{r} 11\frac{10}{12} \\ -\quad 5\frac{9}{12} \\ \hline 6\frac{1}{12} \end{array}$$

The answer is $6\frac{1}{12}$.

Our rules for addition and subtraction of fractions hold in all problems except one kind. That special case is when we are *subtracting* a mixed number from another mixed number, where the fractional part of the mixed number being subtracted is LARGER than the fractional part of the other.

Example 10: $5\frac{3}{4} - 3\frac{7}{8}$

Here we are trying to subtract $\frac{7}{8}$ *from* $\frac{3}{4}$, but $\frac{7}{8}$ is larger than $\frac{3}{4}$.

We can do this problem if we BORROW 1 from the whole number and add it to the fraction part.

We make $5\frac{3}{4} = 4 + 1 + \frac{3}{4} = 4 + 1\frac{3}{4} = 4 + \frac{7}{4}$.

We can now subtract $3\frac{7}{8}$ from $4\frac{7}{4}$.

We proceed in the usual way. We change both fractions to equivalent ones with a new denominator of 8, since 8 is the L.C.M. of 8 and 4.

$$\begin{array}{rcl} 4\frac{7}{4} & \overset{\times 2}{=} & 4\frac{14}{8} \\ -\;3\frac{7}{8} & = & -\;3\frac{7}{8} \\ \hline \end{array}$$

$1\frac{7}{8}$ is the answer after subtracting numerators and number parts.

To review:

$$\begin{array}{rcl} 5\frac{3}{4} = 4 + 1 + \frac{3}{4} & & 4\frac{7}{4} \\ -\;3\frac{7}{8} & = & -\;3\frac{7}{8} \\ \hline & & = 1 + \frac{7}{8} = 1\frac{7}{8} \text{ answer} \end{array}$$

The idea of *borrowing* to do subtraction is tricky. Here is another example to help you see it.

Example 11: $19\frac{2}{5} - 11\frac{7}{8}$

First, you must recognize that you are subtracting a *larger* number from a smaller one. Let us find the L.C.M. of the denominators and the equivalent fraction forms. Then it will become apparent.

The L.C.M. of 5 and 8 is 40. $\frac{2}{5} \overset{\times 8}{=} \frac{16}{40}$ and $\frac{7}{8} \overset{\times 5}{=} \frac{35}{40}$

$$\begin{array}{rcl} 19\frac{2}{5} & & 19\frac{16}{40} \\ -\;11\frac{7}{8} & = & -\;11\frac{35}{40} \\ \hline \end{array}$$

We can now see that $\frac{35}{40}$ is larger than $\frac{16}{40}$. We must borrow 1 from the 19.

Our mixed number becomes $19\frac{16}{40} = 18 + 1 + \frac{16}{40} = 18 + 1\frac{16}{40} = 18\frac{56}{40}$

$$\begin{array}{rcl} 19\frac{2}{5} & & 18\frac{56}{40} \\ -\;11\frac{7}{8} & = & -\;11\frac{35}{40} \\ \hline & & 7\frac{21}{40} \end{array}$$

Our final answer is $7\frac{21}{40}$.

Example 12: Subtract $10 - 7\frac{3}{4}$

Here we are subtracting $\frac{3}{4}$ from nothing. We must borrow so that there is a fraction to subtract from.

The number 10 becomes $9 + 1 = 9\frac{4}{4}$.

$$\begin{array}{rr} 10 & = 9\frac{4}{4} \\ -7\frac{3}{4} & -7\frac{3}{4} \\ \hline & 2\frac{1}{4} \end{array}$$

The answer is $2\frac{1}{4}$.

One last word before you begin the practice exercises. Always remember to REDUCE your answers to LOWEST TERMS.

Good Luck!

ADDITION AND SUBTRACTION OF FRACTIONS EXERCISES

Practice Set A

All answers should be proper fractions or mixed numbers, reduced to lowest terms.

Find the following sums:

1) $\frac{2}{9} + \frac{4}{9}$

2) $\frac{5}{11} + \frac{6}{11}$

3) $\frac{4}{19} + \frac{3}{19} + \frac{7}{19}$

4) $\frac{7}{8} + \frac{5}{8}$

5) $\frac{1}{3} + \frac{4}{3}$

6) $\frac{5}{6} + \frac{1}{6} + \frac{7}{6}$

7) $\frac{3}{4} + \frac{2}{4}$

8) $\frac{1}{2} + \frac{3}{4}$

9) $\frac{2}{3} + \frac{1}{2}$

10) $\frac{7}{8} + \frac{5}{6}$

11) $\frac{4}{3} + \frac{5}{6} + \frac{7}{12}$

12) $\frac{9}{10} + \frac{3}{8}$

13) $\frac{2}{3} + \frac{5}{8} + \frac{4}{9}$

14) $\frac{5}{11} + \frac{7}{22}$

15) $2\frac{2}{3} + 4\frac{1}{3}$

16) $1\frac{3}{5} + 3\frac{7}{10}$

17) $3\frac{1}{2} + 2\frac{1}{4}$

18) $3\frac{3}{4} + 7\frac{1}{2}$

19) $3\frac{5}{8} + 2\frac{3}{4}$

20) $3\frac{5}{6} + 12\frac{1}{3}$

21) $10\frac{7}{8} + 13\frac{3}{4}$

22) $4\frac{2}{3} + 1\frac{1}{2} + 2\frac{1}{4}$

23) $15\frac{2}{5} + 2\frac{1}{2} + 13\frac{2}{3}$

24) $16 + 10\frac{1}{2} + 12\frac{4}{5}$

25) $13\frac{3}{8} + 20\frac{1}{2} + 15\frac{3}{4}$

Find the following differences:

26) $\begin{array}{r} \frac{3}{4} \\ -\frac{2}{4} \\ \hline \end{array}$ 27) $\begin{array}{r} \frac{11}{9} \\ -\frac{7}{9} \\ \hline \end{array}$ 28) $\begin{array}{r} \frac{5}{3} \\ -\frac{2}{3} \\ \hline \end{array}$ 29) $\begin{array}{r} \frac{7}{12} \\ -\frac{5}{12} \\ \hline \end{array}$ 30) $\begin{array}{r} \frac{19}{10} \\ -\frac{7}{10} \\ \hline \end{array}$

31) $\begin{array}{r} \frac{3}{4} \\ -\frac{2}{3} \\ \hline \end{array}$ 32) $\begin{array}{r} \frac{7}{8} \\ -\frac{5}{6} \\ \hline \end{array}$ 33) $\begin{array}{r} \frac{7}{8} \\ -\frac{3}{4} \\ \hline \end{array}$ 34) $\begin{array}{r} \frac{9}{10} \\ -\frac{5}{6} \\ \hline \end{array}$ 35) $\begin{array}{r} 16\frac{3}{4} \\ -\ 7\frac{1}{4} \\ \hline \end{array}$

36) $\begin{array}{r} 23\frac{7}{8} \\ -16\frac{3}{8} \\ \hline \end{array}$ 37) $\begin{array}{r} 12\frac{1}{2} \\ -10 \\ \hline \end{array}$ 38) $\begin{array}{r} 9\frac{3}{4} \\ -3\frac{1}{2} \\ \hline \end{array}$ 39) $\begin{array}{r} 9\frac{3}{4} \\ -3\frac{7}{8} \\ \hline \end{array}$ 40) $\begin{array}{r} 17\frac{7}{8} \\ -11\frac{1}{2} \\ \hline \end{array}$

41) $\begin{array}{r} 19\frac{2}{3} \\ -10\frac{1}{2} \\ \hline \end{array}$ 42) $\begin{array}{r} 100\frac{5}{6} \\ -\ 69\frac{4}{15} \\ \hline \end{array}$ 43) $\begin{array}{r} 12 \\ -10\frac{1}{2} \\ \hline \end{array}$ 44) $\begin{array}{r} 28\frac{1}{2} \\ -19\frac{7}{8} \\ \hline \end{array}$ 45) $\begin{array}{r} 107\frac{1}{2} \\ -\ 99\frac{4}{5} \\ \hline \end{array}$

46) $\begin{array}{r} 96\frac{1}{4} \\ -28\frac{7}{12} \\ \hline \end{array}$ 47) $\begin{array}{r} 15\frac{1}{8} \\ -10\frac{3}{5} \\ \hline \end{array}$ 48) $\begin{array}{r} 105\frac{1}{2} \\ -\ 98\frac{4}{5} \\ \hline \end{array}$ 49) $\begin{array}{r} 5\frac{3}{8} \\ -2\frac{7}{8} \\ \hline \end{array}$ 50) $\begin{array}{r} 25\frac{7}{10} \\ -20\frac{4}{5} \\ \hline \end{array}$

Add and Subtract as indicated in the exercise:

51) $\frac{5}{2} + \frac{3}{2} - \frac{1}{2}$

52) $\frac{7}{8} + \frac{5}{6} - \frac{3}{4}$

53) $12\frac{2}{3} - 10\frac{1}{3} + 4\frac{1}{3}$

54) $7\frac{1}{2} - 5\frac{3}{4} + 1\frac{2}{3}$

55) $7\frac{1}{2} - (5\frac{3}{4} + 1\frac{2}{3})$

56) $\frac{1}{2} - \frac{1}{4} - \frac{1}{8}$

ADDITION AND SUBTRACTION OF FRACTIONS EXERCISES

Practice Set B

Find the following sums:

1) $\frac{4}{7} + \frac{2}{7}$

2) $\frac{6}{13} + \frac{7}{13}$

3) $\frac{5}{12} + \frac{11}{12}$

4) $\frac{2}{3} + \frac{7}{3}$

5) $\frac{7}{8} + \frac{3}{8} + \frac{5}{8}$

6) $\frac{3}{10} + \frac{9}{10}$

7) $\frac{5}{6} + \frac{3}{10}$

8) $\frac{6}{7} + \frac{3}{14}$

9) $\frac{3}{8} + \frac{1}{4}$

10) $\frac{3}{5} + \frac{1}{2}$

11) $\frac{5}{3} + \frac{3}{5} + \frac{7}{15}$

12) $\frac{5}{18} + \frac{7}{12}$

13) $\frac{9}{16} + \frac{5}{8} + \frac{2}{5}$

14) $\frac{31}{100} + \frac{21}{50}$

15) $3\frac{5}{8} + 4\frac{1}{8}$

16) $9\frac{5}{12} + 11\frac{11}{12}$

17) $8\frac{2}{3} + 5\frac{1}{2}$

18) $7\frac{3}{4} + 7\frac{7}{8}$

19) $3\frac{9}{10} + 9\frac{3}{5}$

20) $17\frac{1}{3} + 12\frac{2}{5}$

21) $29\frac{5}{6} + 15\frac{4}{5}$

22) $11\frac{2}{3} + 9\frac{3}{4} + 12\frac{7}{8}$

23) $35\frac{4}{5} + 42\frac{1}{3} + \frac{1}{2}$

24) $17 + 18\frac{4}{9} + 6\frac{1}{4}$

25) $95\frac{1}{8} + 132\frac{3}{4} + 9\frac{2}{9}$

Find the following differences:

26) $\frac{5}{8} - \frac{3}{8}$

27) $\frac{14}{13} - \frac{9}{13}$

28) $\frac{11}{6} - \frac{5}{6}$

29) $\frac{13}{18} - \frac{7}{18}$

30) $\frac{73}{10} - \frac{59}{10}$

31) $\frac{5}{6} - \frac{3}{4}$

32) $\frac{9}{10} - \frac{7}{8}$

33) $\frac{7}{8} - \frac{5}{6}$

34) $18\frac{3}{5} - 11\frac{1}{5}$

35) $21\frac{7}{12} - 13\frac{1}{12}$

36) $85\frac{2}{3} - 75\frac{1}{2}$

37) $25\frac{7}{8} - 19$

38) $25 - 19\frac{7}{8}$

39) $11\frac{2}{3} - 9\frac{4}{5}$

40) $11\frac{2}{3} - 9\frac{3}{4}$

41) $28\frac{7}{10} - 19\frac{4}{5}$

42) $221\frac{5}{6} - 176\frac{3}{5}$

43) $11\frac{3}{4} - 8\frac{5}{4}$

44) $100 - 98\frac{11}{24}$

45) $2\frac{3}{11} - 1\frac{9}{11}$

Add and Subtract as indicated:

46) $\frac{2}{5} + \frac{4}{5} - \frac{3}{5}$

47) $\frac{9}{10} + \frac{7}{8} - \frac{5}{6}$

48) $18\frac{5}{6} - 17\frac{1}{6} + 4\frac{5}{6}$

49) $11\frac{3}{4} - 9\frac{7}{8} + 1\frac{1}{2}$

50) $\frac{2}{3} - \frac{2}{9} - \frac{2}{27}$

APPLICATIONS USING ADDITION AND SUBTRACTION OF FRACTIONS

Practice Set A

1) A bread recipe calls for $8\frac{1}{2}$ pounds of white flour, $2\frac{1}{4}$ pounds of rye flour and $6\frac{3}{4}$ pounds of whole wheat flour. Find the total number of pounds of flour called for in the recipe.

2) Elane and Tina were training to run a half-marathon. One week they ran $5\frac{1}{2}$ miles, then $8\frac{1}{3}$ miles and $10\frac{5}{6}$ miles and then $5\frac{1}{2}$ miles and $10\frac{5}{6}$ miles. How many miles did they run altogether?

3) On a recent trip to San Diego we stopped four times for gas, using $8\frac{2}{5}$ gallons, and $7\frac{1}{2}$ gallons and $10\frac{3}{4}$ gallons and $12\frac{9}{10}$ gallons. Find the total number of gallons purchased.

4) This paper is $8\frac{1}{2}$ by 11 inches. How much would have to be added so that the width is the same as the length?

5) How much larger is $3\frac{5}{4}$ than $3\frac{4}{5}$?

6) A man must fence a field in the shape shown. He has a roll of fencing 150 feet long. How much more does he need, or how much does he have left over?

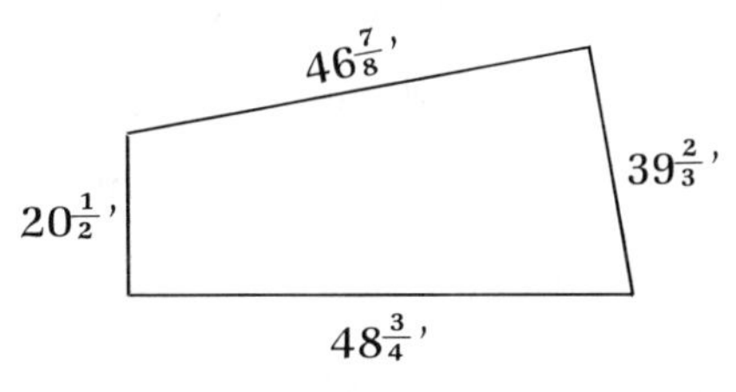

7) A businesswoman found that the cost to market her product was distributed as follows: $13\frac{1}{4}$% to utilities, $29\frac{1}{2}$% to materials, $21\frac{2}{3}$% to labor, and the rest to miscellaneous items. Find the amount of percent of cost attributed to miscellaneous items.

8) A car's gas tank holds $14\frac{1}{2}$ gallons of gas. If it took $8\frac{7}{8}$ gallons to fill it, how many gallons were in the tank before it was filled?

9) A dress designer is constructing a sample of her latest creation. It is a skirt and matching top to be made. The skirt requires $2\frac{3}{4}$ yards of fabric, and the top requires $3\frac{1}{8}$ yards. How much material is needed altogether?

10) You are measuring your living room for building plans. One wall measures $56\frac{1}{4}$ inches to a window, then the window measures $74\frac{1}{8}$ inches. The total of the wall is $162\frac{3}{4}$". What is the remaining measurement?

11) Clothing sizes mean different things depending on what article of clothing. Glove sizes for women were taken when the hand was closed into a fist. It was the measurement around the palm and fingers. A lady's hat size measures the number of inches around her head. However, a man's hat size measures the *diameter* of his head. But a man who blocked hats couldn't figure fractions accurately, and made a mistake. Ever since then, the measurement is $\frac{5}{32}$ of an inch LARGER than the actual hat size.

If a man's hat size is $7\frac{1}{8}$, what is the actual diameter of his head?

12) On Monday, you bought some shares of Compu-Vision stock at $10\frac{3}{8}$ per share (actually represents \$$10\frac{3}{8}$ per share). On Tuesday, it was up $\frac{1}{2}$ point. It was also up on Wednesday by $\frac{1}{4}$ point. Very bad news about the company's earnings, and Thursday it dropped $1\frac{3}{4}$ points. By Friday it edged up $\frac{1}{2}$ point to offset the low drop.

Did you lose or gain on the stock by the end of the week? What was the net change for the week?

13) Minnie the Muncher went on a special diet program for a month. The diet was to last for exactly 4 weeks. The first week she lost $5\frac{1}{2}$ pounds. However, by week 2 her body had stabilized and she lost only $\frac{1}{4}$ pound. Discouraged by that small loss, she gorged out the third week and gained $1\frac{1}{8}$ pounds. But, by the last week she geared up and lost $1\frac{3}{4}$ pounds.

If she weighed $148\frac{1}{8}$ pounds when she started the diet, how much did she weigh at the end of the month?

14) The tallest living man is from Pakistan, and has a height of $98\frac{3}{4}$ inches. The smallest man living is from Portugal and stands $29\frac{1}{2}$ inches tall. If the two were standing side by side, what would be the difference in their heights?

15) You see here a floor plan for a room in a house. Many of the measurements are given. You have enough information to determine the values of the lengths indicated on the plan by the letters A, B, C and D.

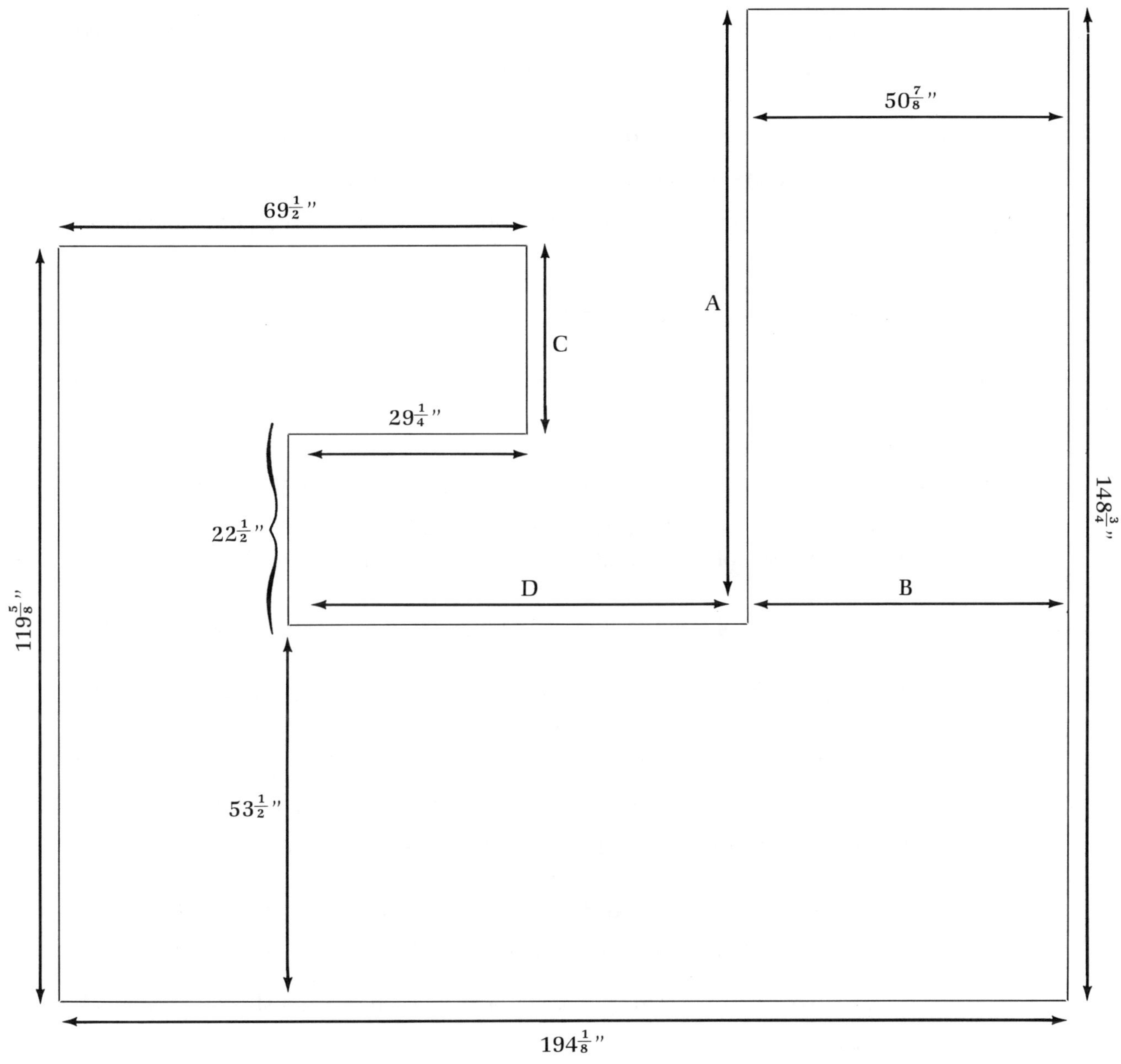

A = __________ B = __________ C = __________ D = __________

APPLICATIONS USING ADDITION AND SUBTRACTION OF FRACTIONS

Practice Set B

1) You paid $\frac{1}{4}$ of a debt in January, $\frac{2}{5}$ of the debt in February and $\frac{1}{10}$ of the debt in March. What portion of the debt do you have *left* to pay?

2) Cassie worked $7\frac{1}{2}$ hours on Monday, $6\frac{2}{3}$ hours on Tuesday, $8\frac{3}{4}$ hours on Wednesday, 9 hours on Thursday, and $6\frac{1}{3}$ hours on Friday. She is supposed to work 40 hours. Was she short some hours or does she need to get paid for overtime? By how many hours?

3) How much larger is $4\frac{4}{3}$ than $4\frac{3}{4}$?

4) Your car's gas tank holds $14\frac{1}{4}$ gallons of gas. If it took $9\frac{7}{8}$ gallons to fill it, how many gallons were in the tank before it was filled?

5) The length of a rectangular field is $156\frac{2}{3}$ yards. The width of the field is $63\frac{7}{8}$ yds.
   a) How much longer is the field than it is wide?
   b) How many feet of fence material is needed to enclose the field?

6) You are constructing several items from a piece of cloth that measures 54 inches wide and 144 inches long. You need to cut out 5 identical patterns that measure 54 inches wide and $26\frac{3}{4}$ inches long. How much material will be *left over*?

7) Each month, $\frac{1}{3}$ of your salary goes to pay for rent, while $\frac{2}{5}$ of your salary pays for the groceries and utilities, and $\frac{1}{4}$ of your salary pays your other monthly bills. What part of your salary is left over for your own enjoyment?

8) You purchased several hundred shares of a certain Silicon Valley company at a price of $\$19\frac{7}{8}$ per share. The activity on the stock looked like this after 1 week: up $1\frac{3}{4}$, down $\frac{1}{2}$, down $\frac{1}{4}$, up $\frac{1}{8}$, up $\frac{7}{8}$. What was the price of each share of the stock by the week's end?

9) You need to buy molding to finish off a rectangular window in your living room. The dimensions of the window are $46\frac{2}{3}$ inches by $37\frac{1}{8}$ inches. Assuming no waste, how long a piece of molding must you buy to completely surround the window?

10) How close is $\frac{4}{15}$ to $\frac{1}{4}$?

SOME CHALLENGING FRACTION PROBLEMS AND PATTERNS

These exercises are a bit more challenging and the results are interesting if you follow the patterns.

1) There is an interesting pattern in this simple series of fractions:

   a) $\frac{1}{2} + \frac{1}{4} =$ ____________

   b) $\frac{1}{2} + \frac{1}{4} + \frac{1}{8} =$ ____________

   c) $\frac{1}{2} + \frac{1}{4} + \frac{1}{8} + \frac{1}{16} =$ ____________

   d) $\frac{1}{2} + \frac{1}{4} + \frac{1}{8} + \frac{1}{16} + \frac{1}{32} =$ ____________

   e) Write out the next three lines of the pattern, and do the additions.

   f) What is the special pattern in the answers?

2) Here is another fraction series with a pattern.

   a) $\frac{1}{3} + \frac{1}{9} =$ ____________

   b) $\frac{1}{3} + \frac{1}{9} + \frac{1}{27} =$ ____________

   c) $\frac{1}{3} + \frac{1}{9} + \frac{1}{27} + \frac{1}{81} =$ ____________

   d) Write out the next two lines of the pattern and do the additions.

   e) What pattern do the answers follow?

3) What is a third and a half of a third and a half of ten?

4) If a watermelon would balance on a scale $\frac{2}{3}$ pounds and $\frac{2}{3}$ of a watermelon, what is the exact weight of 1 watermelon in pounds?

5) Here is another interesting number pattern. It is a very special case and doesn't work for all fractions.

$$1 \times \frac{1}{2} = 1 - \frac{1}{2} = \frac{1}{2}$$

$$2 \times \frac{2}{3} = 2 - \frac{2}{3} = 1\frac{1}{3}$$

$$3 \times \frac{3}{4} = 3 - \frac{3}{4} = 2\frac{1}{4}$$

a) Write out the next five lines of the pattern. Be sure to do the subtraction computations in each part.

b) Fill in the missing numbers in this 13th line of the pattern.

$$13 \times \frac{13}{?} = 13 - \frac{13}{?} = 12\frac{1}{?}$$

c) What is the general pattern? Try to verbalize it.

d) WHY does the pattern work?

6) Take the three consecutive numbers 1, 2, and 3. Using each of these numbers only once in each fraction, FORM the SIX fractions that can be created from these 3 numbers. What is their sum?

7) A fly is at the bottom of a bottle $3\frac{1}{2}$ inches deep. Each day he climbs up $1\frac{1}{4}$ inch, and each night he slips back down $\frac{1}{2}$ inch. How many hours will it take for him to climb out of the bottle? (Assume that day is 12 hours and night is 12 hours.)

MULTIPLICATION AND DIVISION OF FRACTIONS

Multiplying fractions just involves multiplying numerators together, and then multiplying denominators together. The only complication is remembering to put final answers in reduced form.

Example 1: Multiply $\frac{1}{2} \times \frac{3}{4}$

We simply multiply the numerators across, and then multiply the denominators.

$$\frac{1}{2} \times \frac{3}{4} = \frac{1 \times 3}{2 \times 4} = \frac{3}{8}$$

Sometimes we multiply fractions BEFORE reducing them. Then we reduce the answer.

Example 2: Multiply $\frac{2}{3} \times \frac{9}{8}$

Multiply the numerators and then the denominators.

$$\frac{2}{3} \times \frac{9}{8} = \frac{2 \times 9}{3 \times 8} = \frac{18}{24}$$

Now we must reduce the result. $\frac{18}{24} \overset{\div 6}{=} \frac{3}{4}$ (÷ 6)

Many times it is easier to REDUCE first, then multiply. In multiplication exercises, we can reduce by dividing any one numerator and any one denominator in the problem by the SAME number.

Example 3: Multiply $\frac{2}{3} \times \frac{9}{8}$ by *reducing first*.

$$\frac{2}{3} \times \frac{9}{8} = \frac{\cancel{2}^{1}}{\cancel{3}_{1}} \times \frac{\cancel{9}^{3}}{\cancel{8}_{4}} = \frac{1 \times 3}{1 \times 4} = \frac{3}{4}$$

We simply divided the number 2 into both the numerator 2 and the denominator 8. Then we divided 3 into both the denominator 3 and the numerator 9. Then we multiply the new numerators and the new denominators.

It is usually easier to reduce (cancel) first, and then multiply out the numerators and denominators.

Example 3: Multiply $\frac{8}{12} \times \frac{5}{15}$ by reducing first.

First we will reduce the $\frac{8}{12}$ and $\frac{5}{15}$

$$\frac{8}{12} \times \frac{5}{15} = \frac{\cancel{8}^{\textcircled{2}}}{\cancel{12}_{\textcircled{3}}} \times \frac{\cancel{5}^{\textcircled{1}}}{\cancel{15}_{\textcircled{3}}}$$

Then we can multiply the new tops and new bottoms separately.

$$\frac{2 \times 1}{3 \times 3} = \frac{2}{9}$$

Sometimes there are exercises which involve reducing fractions several times. There may be several first steps that can be taken. The order of reducing doesn't matter, as long as when you are finished, there are *no* factors common between any numerator and any denominator. You can reduce "crosswise" or "top to bottom".

Example 4: Multiply $\frac{8}{12} \times \frac{3}{24}$ several ways.

A) Reducing "crosswise" we divide 8 into both 8 and 24. Then we divide the 3 into both 3 and 12.

$$\frac{8}{12} \times \frac{3}{24} = \frac{\cancel{8}^{\textcircled{1}}}{\cancel{12}_{\textcircled{4}}} \times \frac{\cancel{3}^{\textcircled{1}}}{\cancel{24}_{\textcircled{3}}} = \frac{1 \times 1}{4 \times 3}$$

Multiplying the tops and bottoms, we get $\frac{1}{12}$.

$$\frac{1 \times 1}{4 \times 3} = \frac{1}{12}$$

B) Reducing "top to bottom" we divide 4 into both the 8 and the 12. Then we reduce "top to bottom" the 3 into both the 3 and the 24.

$$\frac{8}{12} \times \frac{3}{24} = \frac{\cancel{8}^{\textcircled{2}}}{\cancel{24}_{\textcircled{3}}} \times \frac{\cancel{3}^{\textcircled{1}}}{\cancel{24}_{\textcircled{8}}}$$

Then we cross cancel the 2 with the 8.

$$\frac{8}{12} \times \frac{3}{24} = \frac{\cancel{8}^{\textcircled{1}}}{\cancel{12}_{\textcircled{3}}} \times \frac{\cancel{3}^{\textcircled{1}}}{\cancel{24}_{\textcircled{4}}} = \frac{1 \times 1}{3 \times 4} = \frac{1}{12}$$

The result is still $\frac{1}{12}$.

There are still other orders that we could follow with this problem.

We can perform multiplication of mixed numbers in the same way, if we first change the mixed numbers to fractions.

Example 5: Multiply $2\frac{3}{4} \times 1\frac{1}{3}$

To do this problem we first change all the mixed numbers to improper fractions.

$$2\frac{3}{4} = \frac{2 \times 4 + 3}{4} = \frac{11}{4} \quad \text{and} \quad 1\frac{1}{3} = \frac{1 \times 3 + 1}{3} = \frac{4}{3}$$

Our problem becomes: $\frac{11}{4} \times \frac{4}{3}$.

Now we proceed to cancel "crosswise".

$$\frac{11}{\cancel{4}_{1}} \times \frac{\cancel{4}^{1}}{3} = \frac{11 \times 1}{1 \times 3} = \frac{11}{3}$$

Our answer is $\frac{11}{3}$ or $3\frac{2}{3}$.

Example 6: Multiply $\frac{2}{3} \times 24$

The whole number 24 must be put in fraction form. Any whole number "n" can be expressed as $\frac{n}{1}$.

Our problem becomes $\frac{2}{3} \times \frac{24}{1}$. We work this by cross cancelling.

$$\frac{2}{3} \times \frac{24}{1} = \frac{2}{\cancel{3}_{1}} \times \frac{\cancel{24}^{8}}{1} = \frac{2 \times 8}{1 \times 1} = \frac{16}{1} = 16$$

The answer is 16.

Dividing fractions is as simple as multiplying fractions. The only twist is to INVERT any fraction that follows the division sign (÷). After inverting, we proceed to multiply the fractions. We must remember to reduce wherever possible.

Example 7: Divide: $\frac{2}{3} \div \frac{4}{9}$

We invert the $\frac{4}{9}$ and then we have the multiplication problem

$$\frac{2}{3} \times \frac{9}{4}$$

We then proceed to work this problem as a multiplication exercise.

$$\frac{2}{3} \times \frac{9}{4} = \frac{\cancel{2}^{1}}{\cancel{3}_{1}} \times \frac{\cancel{9}^{3}}{\cancel{4}_{2}} = \frac{1 \times 3}{1 \times 2} = \frac{3}{2}$$

Example 8: Divide: $\frac{3}{2} \div \frac{5}{2} \div \frac{9}{10}$

Here we must invert BOTH the fractions after the ÷. We invert both the $\frac{5}{2}$ and the $\frac{9}{10}$. We then have the multiplication problem

$$\frac{3}{2} \times \frac{2}{5} \times \frac{10}{9}.$$

We then work the problem as a usual multiplication exercise.

$$\frac{3}{2} \times \frac{2}{5} \times \frac{10}{9} = \frac{\overset{①}{\cancel{3}}}{\underset{①}{\cancel{2}}} \times \frac{\overset{①}{\cancel{2}}}{\underset{①}{\cancel{5}}} \times \frac{\overset{②}{\cancel{10}}}{\underset{③}{\cancel{9}}} = \frac{1 \times 1 \times 2}{1 \times 1 \times 3} = \frac{2}{3}$$

The answer is $\frac{2}{3}$.

Example 9: Divide: $3\frac{1}{2} \div 14$

Here we need to change the mixed number $3\frac{1}{2}$ to fraction form. $(3\frac{1}{2} = \frac{7}{2})$

Then we change the whole number 14 to a fraction. $(14 = \frac{14}{1})$

Our division problem is $\frac{7}{2} \div \frac{14}{1}$.

NOW we can INVERT the $\frac{14}{1}$ and get a new multiplication problem.

Our problem is $\frac{7}{2} \times \frac{1}{14}$.

Working the fractions we get $\frac{1}{4}$ as the answer.

$$\frac{7}{2} \times \frac{1}{14} = \frac{\overset{①}{\cancel{7}}}{2} \times \frac{1}{\underset{②}{\cancel{14}}} = \frac{1 \times 1}{2 \times 2} = \frac{1}{4}$$

MULTIPLICATION AND DIVISION OF FRACTIONS

Practice Set A

Perform the following multiplication computations.

1) $\frac{3}{4} \times \frac{8}{9}$

2) $\frac{4}{5} \times \frac{15}{8}$

3) $\frac{2}{3} \times \frac{4}{5}$

4) $\frac{5}{8} \times \frac{3}{10}$

5) $\frac{3}{5} \times \frac{4}{5}$

6) $\frac{5}{8} \times \frac{8}{9}$

7) $\frac{7}{8} \times \frac{4}{9}$

8) $\frac{2}{3} \times \frac{9}{10}$

9) $12 \times \frac{3}{4}$

10) $\frac{5}{6} \times 18$

11) $\frac{3}{4} \times \frac{7}{9} \times \frac{3}{14}$

12) $24 \times \frac{2}{3}$

13) $\frac{7}{8} \times 1\frac{1}{3}$

14) $\frac{9}{4} \times 1\frac{5}{27}$

15) $(\frac{1}{2})^3$

16) $18 \times \frac{2}{3} \times \frac{2}{9}$

17) $8\frac{1}{4} \times 1\frac{1}{3}$

18) $\frac{1}{2} \times \frac{1}{3} \times \frac{1}{4}$

19) $2 \times 3\frac{1}{4}$

20) $2\frac{1}{2} \times 5$

21) $66\frac{2}{3} \times 3$

22) $2\frac{1}{3} \times \frac{3}{8}$

23) $(1\frac{2}{3})^2$

24) $1\frac{2}{3} \times 2\frac{3}{4} \times \frac{9}{10}$

25) $1\frac{1}{2} \times \frac{1}{2}$

26) $1\frac{1}{2} \times 2\frac{1}{2}$

27) $1\frac{1}{2} \times 1\frac{1}{2}$

28) $(\frac{2}{3})^3$

29) $1\frac{1}{2} \times \frac{2}{3}$

30) $1\frac{4}{11} \times 1\frac{8}{25}$

Very often the word "of" is used with fractions and means "multiply". For example, $\frac{4}{5}$ of 20 means $\frac{4}{5} \times 20$. Find:

31) $\frac{4}{5}$ of 10

32) $\frac{3}{4}$ of 8

33) $\frac{2}{3}$ of 33

34) $\frac{1}{2}$ of $\frac{1}{2}$ of $\frac{1}{2}$

35) $\frac{5}{6}$ of 42

Perform the following division exercises:

36) $\frac{3}{4} \div \frac{1}{8}$

37) $\frac{1}{2} \div \frac{1}{2}$

38) $\frac{3}{2} \div \frac{2}{3}$

39) $\frac{5}{16} \div \frac{15}{8}$

40) $\frac{7}{8} \div \frac{1}{3}$

41) $\frac{7}{8} \div 3$

42) $100 \div \frac{20}{3}$

43) $6 \div \frac{2}{3}$

44) $\frac{2}{3} \div 6$

45) $2 \div \frac{1}{2}$

46) $\frac{1}{2} \div 2$

47) $\frac{1}{2} \div \frac{1}{2} \div \frac{1}{2}$

48) $\frac{1}{3} \div 3$

49) $12 \div 3 \div 5$

50) $7 \div 10$

51) $34 \div 51$

52) $2\frac{1}{8} \div 4\frac{1}{4}$

53) $3\frac{3}{4} \div 5$

54) $2\frac{5}{8} \div 2\frac{3}{4} \div \frac{7}{11}$

55) $1 \div 2 \div 3$

56) $8 \div \frac{1}{2} \div \frac{4}{3}$

57) $\dfrac{4}{\frac{1}{2}}$

58) $\frac{1}{3} \div \frac{1}{3} \div \frac{1}{3}$

59) $9\frac{9}{10} \div 2\frac{1}{5}$

60) $\dfrac{\frac{9}{10}}{\frac{3}{2}}$

When we INVERT a fraction, we find a number that is the *RECIPROCAL* of the given fraction. For example: the reciprocal of $\frac{2}{3}$ is $\frac{3}{2}$. The reciprocal of $1\frac{3}{4}$ $(=\frac{7}{4})$ is $\frac{4}{7}$. Find the *reciprocal* of the following fractions.

61) $\frac{3}{4}$

62) $\frac{5}{6}$

63) $\frac{1}{4}$

64) $2\frac{1}{2}$

65) $5\frac{3}{10}$

66) $\frac{85}{19}$

67) 5

68) 2

69) $\frac{1}{10}$

70)* $\frac{0}{2}$

Perform the following computations. Work the parentheses first.

71) $(\frac{3}{4} + \frac{1}{3}) \div (\frac{2}{3} + \frac{5}{12})$

72) $(\frac{1}{2} + \frac{1}{4})^2$

73) $2 - (\frac{2}{3} \times \frac{9}{10})$

74) $(2 - \frac{2}{3}) \times \frac{9}{10}$

75) $1\frac{1}{2} - (\frac{2}{3})^2$

MULTIPLICATION AND DIVISION OF FRACTIONS

Practice Set B

Perform the following multiplication computations:

1) $\frac{4}{5} \times \frac{15}{28}$

2) $\frac{3}{4} \times \frac{12}{21}$

3) $\frac{2}{5} \times \frac{4}{3}$

4) $\frac{3}{8} \times \frac{10}{9}$

5) $\frac{1}{3} \times \frac{4}{3}$

6) $\frac{4}{7} \times \frac{7}{8}$

7) $\frac{5}{2} \times \frac{4}{25}$

8) $\frac{5}{12} \times \frac{3}{5}$

9) $15 \times \frac{4}{5}$

10) $\frac{2}{3} \times 21$

11) $\frac{4}{7} \times \frac{9}{12} \times \frac{14}{27}$

12) $56 \times \frac{3}{14}$

13) $\frac{7}{8} \times 3\frac{1}{5}$

14) $\frac{5}{12} \times 20\frac{4}{7}$

15) $(\frac{2}{3})^2$

16) $24 \times \frac{3}{8} \times \frac{5}{6}$

17) $15\frac{5}{6} \times 2\frac{2}{5}$

18) $\frac{2}{3} \times \frac{3}{4} \times \frac{1}{5}$

19) $11 \times 1\frac{1}{22}$

20) $4\frac{1}{2} \times 6$

21) $2\frac{2}{3} \times 3\frac{1}{10}$

22) $7\frac{1}{3} \times 7\frac{1}{2}$

23) $6\frac{3}{10} \times 10$

24) $6\frac{1}{5} \times 7\frac{2}{3}$

25) $1\frac{2}{3} \times 2\frac{2}{3}$

26) $(5\frac{1}{4})^2$

27) $(\frac{1}{2})^4$

28) $2\frac{1}{3} \times 1\frac{1}{2}$

29) $(1\frac{1}{2})^3$

Very often the word "of" is used with fractions and means "multiply".
For example: $\frac{2}{3}$ of 18 means $\frac{2}{3} \times 18$. Find:

30) $\frac{7}{8}$ of 24

31) $\frac{3}{5}$ of 45

32) $\frac{1}{2}$ of $\frac{1}{3}$ of $\frac{1}{4}$

33) $\frac{1}{3}$ of $\frac{1}{3}$ of $\frac{1}{3}$

34) $\frac{9}{10}$ of 125

Perform the following division exercises:

35) $\frac{7}{8} \div \frac{3}{4}$

36) $\frac{2}{3} \div \frac{3}{2}$

37) $\frac{3}{2} \div \frac{3}{2}$

38) $\frac{5}{12} \div \frac{5}{9}$

39) $\frac{5}{6} \div \frac{1}{3}$

40) $\frac{5}{6} \div 3$

41) $50 \div \frac{4}{3}$

42) $12 \div \frac{3}{2}$

43) $\frac{3}{2} \div 12$

44) $9 \div \frac{1}{3}$

45) $\frac{1}{3} \div 9$

46) $\frac{2}{3} \div \frac{2}{3} \div \frac{2}{3}$

47) $\frac{4}{5} \div 4$

48) $18 \div 3 \div \frac{1}{2}$

49) $11 \div 10$

50) $65 \div 91$

51) $8\frac{1}{6} \div 1\frac{1}{6}$

52) $7\frac{1}{3} \div 4$

53) $14\frac{1}{6} \div 4\frac{1}{5} \div \frac{17}{18}$

54) $2 \div 3 \div 4$

55) $9 \div \frac{3}{2} \div 1\frac{1}{5}$

56) $\frac{\frac{3}{2}}{5}$

57) $5 \div \frac{1}{5} \div 5$

58) $10\frac{1}{4} \div 4\frac{5}{8}$

59) $\frac{\frac{7}{8}}{2\frac{11}{12}}$

When we invert a fraction, we find a number that is the *RECIPROCAL* of the given fraction. For example: the reciprocal of $\frac{4}{5}$ is $\frac{5}{4}$. The reciprocal of $1\frac{1}{2}$ $(= \frac{3}{2})$ is $\frac{2}{3}$. Find the reciprocal of the following fractions:

60) $\frac{7}{8}$

61) $\frac{9}{10}$

62) $\frac{1}{5}$

63) $7\frac{1}{2}$

64) $2\frac{2}{3}$

65) $\frac{65}{14}$

66) 3

67) 11

68) $\frac{1}{12}$

*69) $\frac{0}{3}$

Perform the following computations. Work the parentheses first:

70) $(\frac{2}{9} + \frac{1}{3}) \div (\frac{2}{3} + \frac{1}{4})$

71) $(\frac{2}{3} + \frac{1}{5})^2$

72) $4 - (\frac{3}{4} \times \frac{12}{39})$

73) $(4 - \frac{3}{4}) \times \frac{12}{39}$

74) $2\frac{1}{2} - (\frac{3}{4})^2$

APPLICATIONS OF MULTIPLICATION AND DIVISION OF FRACTIONS:

Practice Set A

1) A student got $\frac{4}{5}$ of all the 120 questions on an exam correct. How many did she *miss?*

2) If a backpacker has covered $\frac{3}{4}$ of a 220-mile trip, how many miles does he have *left* to go?

3) A college instructor earns $32,400 per year. If he expects a raise equivalent to $\frac{3}{100}$ of his present pay, what will be his salary next year?

4) How many bags, each containing $1\frac{1}{2}$ pounds of rice, could be filled with 36 pounds of rice?

5) One inch is approximately $2\frac{1}{2}$ centimeters. Find the number of centimeters in $18\frac{1}{2}$ inches.

6) A case of frying pans weighs $141\frac{1}{2}$ pounds. If each frying pan weighs $5\frac{3}{8}$ lbs., and the box holding them weighs $12\frac{1}{2}$ lbs., how many frying pans are in the box?

7) Suppose your sports car went 395 miles on $19\frac{3}{4}$ gallons of gas. How many miles per gallon did you get on that trip?

8) A store owner bought $34.00 worth of ball point pens. If she paid $8\frac{1}{2}$¢ for each pen, how many did she buy?

9) If a sports car travels 24 miles per gallon of gasoline, how many miles will it travel on $5\frac{1}{2}$ gallons of gasoline?

10) If you drive 105 miles in $1\frac{1}{2}$ hours, how fast are you driving?

11) A kilogram is $2\frac{1}{5}$ pounds approximately. How many kilograms does a 132 pound woman weigh?

12) For outstanding women runners, $\frac{3}{25}$ of their body weight is fat. If an outstanding female distance runner weighs 150 lbs., how many pounds of fat does she have?

13) A gallon of water weighs about $8\frac{1}{4}$ pounds. How many gallons of water are in a tank that is holding 627 pounds of water?

14) A bundle of packaging cartons weighs $11\frac{3}{4}$ pounds. What is the total weight of 10 bundles?

15) A car's gas tank holds $11\frac{1}{4}$ gallons of gasoline. How many gallons are in the gas tank if the tank is $\frac{2}{3}$ full?

16) A superball rebounds on each bounce to $\frac{9}{10}$ of the distance it fell. If it was dropped from a height of 10 feet,

a) how high did it rebound after the first bounce?

b) after the second bounce?

c) after the third bounce?

*d) how many bounces before it is LESS than 6 feet from the ground?

17) Gravity at the surface of the Moon is $\frac{1}{6}$ as great as gravity at the surface of the Earth. If a baby weighs $8\frac{2}{5}$ pounds on Earth, how much would she weigh on the Moon?

18) The probability that a certain experimental drug will be effective is $\frac{2}{5}$. That means that it is expected to cure $\frac{2}{5}$ of the people who take the drug. If 250 patients are treated with the drug, how many would be *expected* to be cured?

19) In a 1982 census taken in the People's Republic of China, there were found to be 3765 centenarians (people at least 100 years old). Two-thirds were found to be women. How many female centenarians are there in China (based on the census)?

20) Muscles in the human body account for $\frac{40}{100}$ of a person's total body weight. How much do your muscles weigh?

APPLICATION OF MULTIPLICATION AND DIVISION OF FRACTIONS EXERCISES:

Practice Set B

1) An employee has $\frac{2}{11}$ of her salary withheld for income tax. If her weekly salary is $\$643\frac{1}{2}$, how much is withheld for income tax?

2) The population of a town was 81,000 in the last census. It has increased $\frac{2}{3}$ since then. What is its present population?

3) A car is run until its gas tank shows that it is $\frac{3}{8}$ full. The tank is then filled to capacity by putting in 15 gallons. What is the total capacity of the gas tank of the car?

4) How many bottles of expensive French perfume holding $\frac{2}{3}$ of an ounce can be filled from a quart of the perfume? (HINT: change 1 quart to ounces)

5) A mile is approximately $1\frac{3}{5}$ kilometers. Find the number of kilometers equivalent to 10 miles.

6) Paula and Ellen are wallpapering their apartment's kitchen. If Paula papers $\frac{1}{4}$ of the room, and Ellen papers $\frac{1}{8}$ of the remainder, what fraction of the kitchen is LEFT to paper?

7) How many $\frac{2}{3}$-foot strips of wood can be cut from a board that is 8 feet long?

8) A semiconductor manufacturer finds that $\frac{1}{24}$ of the computer chips shipped in any order are defective. How many chips should be shipped to a customer to insure that 2070 are good?

9) Each year a car depreciates a certain fraction of its value. Support that during the first year it depreciates $\frac{1}{4}$ of its original value. The second year it depreciates $\frac{1}{5}$ of its year-end value. The third year it depreciates $\frac{1}{6}$ of its second year value. What fraction of its original value is the car after the third year?

*10) In olden days, back in Shakespeare's England, a miller was accustomed to take a toll from each customer. Let us suppose that he took one-ninth of the flour he ground for any customer. How much did he grind for a man who had just 1 bushel of flour left after the toll was taken?

11) At the end of the year a taxpayer is filling out his income tax return. He is entitled to take a deduction for the sales tax on his new automobile. If he lost his sales receipt, but he knows that the tax is $\frac{13}{200}$ ($6\frac{1}{2}$%) of the cost of the car, and that amount was $845, what was the price of the car?

12) A farmer wishes to fence a rectangular area for his horses. If the width is to be $\frac{3}{4}$ the length, and it costs $186 for the wood to fence one width, how much will it cost to fence all 4 sides?

*13) Two brothers and a sister were bequeathed an inheritance when dear Aunt Agnes died. Allan was to get $\frac{1}{3}$ of the inheritance while Bonnie was to get $\frac{2}{5}$ of it, leaving the rest to Clyde. When the estate was finally adjusted, Allan had to give $\frac{3}{5}$ of his share to Clyde. Of that increased amount, Clyde had to give $\frac{3}{7}$ of it to Bonnie. What part of Aunt Agnes' entire inheritance did Bonnie finally get?

*14) A bridge crosses a river which measures 660 feet wide from the edge of one bank to another. One bank holds $\frac{1}{4}$ of the bridge, while the other holds $\frac{1}{5}$ of it. How long is the bridge?

MORE DIVISION OF FRACTIONS EXERCISES

Practice Set A

There is a type of problem that compares one number to another, and that involves division. Usually it looks like this:

$$\text{WHAT PART } \textit{OF } 1\frac{1}{2} \textit{ IS } \frac{3}{4}?$$

To do this problem, we set up a ratio or fraction that looks like this:

$$\frac{\text{IS}}{\text{OF}} = \frac{\frac{3}{4}}{1\frac{1}{2}} = \frac{\frac{3}{4}}{\frac{3}{2}} = \frac{3}{4} \times \frac{2}{3} = \frac{1}{2}$$

The RULE is to set up the fraction where the number associated with IS goes in the numerator, and the number after OF goes in the denominator. Then we proceed to DIVIDE.

$$\frac{\text{IS}}{\text{OF}} = \text{IS} \div \text{OF}$$

Let's try some of these problems:

1) What part of $2\frac{1}{2}$ is $1\frac{1}{4}$

2) What part of $8\frac{3}{4}$ is $\frac{7}{8}$?

3) $\frac{7}{8}$ of what number is $\frac{1}{2}$?

4) What part of $4\frac{1}{2}$ is 3?

5) What fraction of 10 is $1\frac{2}{3}$?

6) What fraction of $1\frac{2}{3}$ is 10?

7) What fraction of 8 is $\frac{1}{2}$?

8) What fraction of $\frac{1}{2}$ is 8?

9) $\frac{3}{4}$ of what number is 5?

10) $3\frac{1}{2}$ times what number is 14?

MORE DIVISION OF FRACTIONS EXERCISES

Practice Set B

1) What part of $3\frac{1}{2}$ is $1\frac{3}{4}$?

2) What part of $9\frac{3}{4}$ is $6\frac{1}{2}$?

3) What part of 5 is $1\frac{1}{2}$?

4) $\frac{2}{3}$ of what number is $\frac{3}{4}$?

5) What fraction of 12 is $2\frac{2}{5}$?

6) What fraction of $2\frac{2}{5}$ is 12?

7) What fraction of 6 is $\frac{1}{3}$?

8) What fraction of $\frac{1}{3}$ is 6?

9) $\frac{7}{8}$ of what number is 7?

10) $5\frac{2}{3}$ of what number is 6?

RATIO AND PROPORTION

Suppose that a class has 15 male students and 25 female students.

The RATIO of male *to* female students is 15:25 (also written as $\frac{15}{25}$).

Since $\frac{15}{25} = \frac{3}{5}$ when reduced, we can say that the ratio is also 3:5 (or $\frac{3}{5}$). That is, for every 3 male students there are 5 female students.

When we set 2 ratios equal, we have a PROPORTION. For example; $\frac{15}{25} = \frac{3}{5}$ is a proportion.

A PROPORTION IS AN EQUATION WHERE TWO RATIOS ARE SET EQUAL. In every proportion, *DIAGONAL PRODUCTS ARE EQUAL.*

For example, in the proportion $\frac{15}{25} = \frac{3}{5}$ the diagonal products are equal.

$$\frac{15}{25} \times \frac{3}{5} \qquad 15 \times 5 = 3 \times 25 = 75$$

Recall from our work with equivalent fractions that this is always true. When two fractions are equal, you can cross multiply and get equal values.

Recall that with $\frac{2}{4} = \frac{1}{2}$, then $2 \times 2 = 1 \times 4$.

$$\frac{2}{4} \times \frac{1}{2} \rightarrow 2 \times 2 = 1 \times 4 = 4$$

We can use this idea to find the missing parts of any proportion.

Example 1: Solve the proportion for the missing number.

$$\frac{1}{5} = \frac{7}{\square}$$

Using the fact that diagonal products are equal,

$1 \times \square = 7 \times 5$

or $\square = 7 \times 5 = 35$.

Our unknown value, $\square$, is 35.

Example 2: Solve the proportion $\frac{4}{12} = \frac{3}{\square}$

Since diagonal products are equal, $\frac{4}{12} \times \frac{3}{\square}$

$4 \times \square = 3 \times 12$

$4 \times \square = 36$ (this sentence asks WHAT number times 4 equals 36?)

To find the missing number $\square$, we DIVIDE by 4.

$\square = 36 \div 4$

Since we were *multiplying* by 4 to get 36, we simply UNDO the multiplication by *dividing* by 4 to get $\square$ alone.

$\square = 36 \div 4 = 9$

The result is correct if we recall that $\frac{4}{12} = \frac{1}{3}$ and $\frac{3}{\boxed{9}} = \frac{1}{3}$.

Example 3: Solve the proportion $\frac{3\frac{1}{2}}{7} = \frac{2\frac{1}{4}}{\square}$

Equating diagonal products, we get: $3\frac{1}{2} \times \square = 2\frac{1}{4} \times 7$.

We must change the mixed numbers to fraction form.

$\frac{7}{2} \times \square = \frac{9}{4} \times 7$

Now, to get the missing value, $\square$, we DIVIDE by $\frac{7}{2}$.

$\square = \frac{9}{4} \times 7 \div \frac{7}{2}$

Computing our answer

$$\square = \frac{9}{4} \times \frac{7}{1} \times \frac{2}{7} = \frac{9}{\underset{2}{\cancel{4}}} \times \frac{\overset{1}{\cancel{7}}}{1} \times \frac{\overset{1}{\cancel{2}}}{\underset{1}{\cancel{7}}} = \frac{9}{2} = 4\frac{1}{2}$$

Intuitively this is correct if we think $\frac{3\frac{1}{2}}{7} = \frac{1}{2}$ and $\frac{2\frac{1}{4}}{4\frac{1}{2}} = \frac{1}{2}$.

In our everyday life we encounter many things that are *proportional*.

For example, if 5 pounds of sugar cost \$1.50, then 10 pounds of sugar costs twice as much, \$3.00.

Setting up the situation as a proportion, we get $\frac{5 \text{ lb.}}{10 \text{ lb.}} = \frac{\$1.50}{\$3.00}$.

The ratios $\frac{5}{10}$ and $\frac{1.50}{3.00}$ are both equal to $\frac{1}{2}$.

We can now proceed to use proportions to give us answers to everyday questions.

Example 4: If a car travels 36 miles on $1\frac{1}{2}$ gallons of gasoline, find out how many miles the car will travel on $5\frac{1}{2}$ gallons.

This is an example of proportional quantities since the number of miles travelled is proportional to the number of gallons used. We will develop an organized approach to arrive at a solution.

Step 1: set up a GRID (this will keep our quantities organized)

# miles	# gallons
36	$1\frac{1}{2}$
□	$5\frac{1}{2}$

We indicate the unknown quantity by an open variable box, □.

Step 2: set up a proportion.

The ratios can be set up either reading DOWN the table like

$$\frac{36}{\square} = \frac{1\frac{1}{2}}{5\frac{1}{2}}$$

Or reading ACROSS the colums like $\frac{36}{1\frac{1}{2}} = \frac{\square}{5\frac{1}{2}}$

Step 3: Get equal diagonal products.

We cross multiply to get $1\frac{1}{2} \times \square = 36 \times 5\frac{1}{2}$

Step 4: Proceed to DIVIDE by $1\frac{1}{2}$ to get the answer.

$\square = 36 \times 5\frac{1}{2} \div 1\frac{1}{2}$

We now change mixed numbers to fraction form, invert $1\frac{1}{2}$, and multiply to get the result.

$$\square = \frac{36}{1} \times \frac{11}{2} \div \frac{3}{2} = \frac{36}{1} \times \frac{11}{2} \times \frac{2}{3} = \frac{\overset{12}{\cancel{36}}}{1} \times \frac{11}{\underset{1}{\cancel{2}}} \times \frac{\overset{1}{\cancel{2}}}{\underset{1}{\cancel{3}}} =$$

$12 \times 11 = 132$

The answer is 132 miles.

Example 5: If $\frac{1}{2}$ cup of dried milk can be mixed with 2 cups water to make a quart of reconstituted milk, how much milk should be mixed with 1 gallon (16 cups) water?

Step 1: set up the GRID

# cups dried milk	# cups water
$\frac{1}{2}$	2
$\square$	16

Step 2: set up the RATIOS

reading down $\dfrac{\frac{1}{2}}{\square} = \dfrac{2}{16}$ or reading across $\dfrac{\frac{1}{2}}{2} = \dfrac{\square}{16}$

Step 3: set diagonal products equal

$$2 \times \square = \frac{1}{2} \times 16$$

Step 4: divide by 2

$$\square = \frac{1}{2} \times 16 \div 2$$

The answer is 4 cups of dried milk.

$$\square = \frac{1}{2} \times 16 \div 2 = \frac{1}{2} \times \frac{16}{1} \times \frac{1}{2} = \frac{16}{4} = 4$$

Example 6: The ratio of body fat to body weight for outstanding female distance runners is approximately 3:25. If an outstanding female distance runner weighs 150 pounds, then how many pounds of fat does she have?

Step 1: set up the GRID

# lbs. body fat	#lbs. body weight
3	25
□	150

The ratio 3:25 means that for every 3 lbs. of body fat there are 25 lbs. of body weight.

Step 2: set up the proportion

$$\frac{3}{\square} = \frac{25}{150}$$

Step 3: equate the diagonal products

$$25 \times \square = 3 \times 150$$

Step 4: divide by 25

$$\square = 3 \times 150 \div 25 = 450 \div 25 = 18$$

The female runner has 18 lbs. of body fat.

Let's review the steps needed to solve any ratio and proportion application.

STEP 1: SET UP THE *GRID*.
STEP 2: SET UP THE PROPORTION (this can be done reading DOWN or ACROSS).
STEP 3: SET THE DIAGONAL PRODUCTS EQUAL.
STEP 4: DIVIDE BY THE NUMBER NEXT TO THE □.

RATIO AND PROPORTION EXERCISES

Practice Set A

Solve the following proportion equations for the missing number:

1) $\frac{11}{33} = \frac{8}{\square}$

2) $\frac{15}{60} = \frac{\square}{100}$

3) $\frac{5}{6} = \frac{\square}{48}$

4) $\frac{2\frac{1}{2}}{1} = \frac{\frac{1}{2}}{\square}$

5) $\frac{4\frac{2}{3}}{3} = \frac{\square}{2}$

6) $\frac{1\frac{3}{4}}{4} = \frac{\square}{12}$

7) $\frac{2\frac{2}{5}}{5} = \frac{6}{\square}$

8) $\frac{6}{\square} = \frac{4\frac{1}{2}}{1\frac{1}{2}}$

9) $\frac{3\frac{1}{8}}{5} = \frac{\square}{2\frac{2}{3}}$

10) $\frac{12\frac{1}{2}}{\frac{5}{3}} = \frac{\square}{1\frac{5}{9}}$

Set up an appropriate grid and answer the following:

11) The scale of a map reads 5 miles for every $\frac{1}{2}$ inch on the map. If the line connecting two points on a map is $4\frac{3}{4}$ inches, how many miles apart are the two points)? (HINT: set up a grid with # inches and # miles as headings)

12) If you know that water weighs approximately $62\frac{1}{2}$ pounds per 1 cubic foot, and your bathtub holds 250 cubic feet of water, how much would the water filling the bathtub weigh? (HINT: set up a grid with # lbs. water and # cubic feet as the headings).

13) For every 3 electoral votes, candidate Smiley gets 49,000 popular votes. If Smiley got 39 electoral votes, how many popular votes did he get? (HINT: set up the grid with the column headings of # electoral votes and # popular votes)

14) A kilogram is equivalent to $2\frac{1}{5}$ pounds. How many kilograms does a $113\frac{3}{10}$ pound woman weigh?

15) If you drive $92\frac{3}{4}$ miles in $1\frac{2}{5}$ hours, at what speed are you driving? (how many miles per 1 hour?)

16) If a $5\frac{1}{2}$ foot woman casts a $3\frac{1}{2}$ foot shadow, how tall is a tree that casts a 14 foot shadow? (Hint: set up a grid with the headings # ft. shadow and # actual ft.)

17) An outstanding female marathon runner can run 8 miles in $\frac{2}{3}$ hours.
   a) How many miles can she run in $1\frac{1}{2}$ hours?
   b) If she lives $22\frac{1}{2}$ miles from a college track field, how long would it take her to run to the field?

18) The adult human brain has a ratio of 76 grams water to 98 grams weight. If a person's brain weighs 147 grams, how many grams of water does the brain have?

19) Three ears of corn indirectly require 216 gallons of water for them to grow. How much water will be consumed to provide a family of 5 with one ear of corn per person for one dinner?

20) Alcohol is known to increase the output of urine. For every 2 ounces of alcohol consumed, 18 ounces of extra urine are produced. How much urine is produced by a person who drinks a fifth of whiskey containing $27\frac{2}{3}$ ounces alcohol?

21) If a sack of coffee weighs 78 pounds when filled and you have $\frac{2}{3}$ of a sack filled, how many pounds of coffee do you have?

22) If $2\frac{1}{2}$ gallons of paint cover a wall area of 100 square feet, how many gallons are needed to paint a space of 240 square feet?

23) If you use 70 therms of gas in 30 days, and you continue to use gas at the same rate, how many therms of gas will you use in 50 days?

24) a) A photograph 4 inches wide and 6 inches long is enlarged proportionally so that it is 14 inches wide. How long is the enlargement?

b) In the 4" × 6" photograph there is a painting that measures 1 in. by $1\frac{1}{2}$ inches. The woman in the photograph measures 3 inches in height, although in reality she is 5 ft. 5 in. tall. In reality, what are the dimensions of the painting?

25) The following recipe only makes enough to serve 6 average portions. List each ingredient, and find how much of each is needed to make 10 average servings:

Ambrosial Strawberries

30 fresh strawberries
3 ounces (6 Tbsp.) tawny port
4 canned figs
3 ounces (6 Tbsp.) kirsch liquor
2 cups sweetened whipped cream
6 scoops vanilla ice cream
4 macaroons
2 teaspoons Grand Marnier

*26) If it takes $4\frac{7}{8}$ minutes to boil one egg, how long will it take to boil 4 eggs?

*27) If a hundred ostrich can eat a hundred pounds of worms in a hundred days, how long will it take ten ostrich to eat ten pounds of worms?

*28) Speaking in averages, if one and a half ostrich can lay one and a half eggs in one and a half days, how many days will it take one ostrich to lay one and a half dozen eggs?

"HOW MUCH FOR ONE, SONNY"

RATIO AND PROPORTION EXERCISES

Practice Set B

Solve the following proportion equations for the missing number:

1) $\frac{12}{36} = \frac{7}{\square}$

2) $\frac{14}{84} = \frac{\square}{150}$

3) $\frac{7}{8} = \frac{\square}{56}$

4) $\frac{3\frac{3}{4}}{1} = \frac{\square}{\frac{2}{3}}$

5) $\frac{3\frac{1}{3}}{\frac{5}{6}} = \frac{\square}{4}$

6) $\frac{2\frac{2}{5}}{6} = \frac{\square}{\frac{1}{2}}$

7) $\frac{100}{12\frac{1}{2}} = \frac{5}{\square}$

8) $\frac{3}{4\frac{1}{2}} = \frac{\square}{6}$

9) $\frac{7}{\square} = \frac{8}{5\frac{1}{7}}$

10) $\frac{\frac{9}{10}}{12} = \frac{3}{\square}$

Set up an appropriate grid and answer the following:

11) The scale of a map reads $4\frac{1}{2}$ miles for every $\frac{3}{4}$ inch on the map. If the line connecting two points on the map is $5\frac{5}{8}$ inches, how many miles apart are the two points?

12) A kilogram is equivalent to $2\frac{2}{10}$ pounds. If a man weighs $157\frac{3}{10}$ pounds, how many kilograms is that?

13) If you drive $113\frac{3}{4}$ miles in $1\frac{3}{4}$ hours, at what speed are you driving? (how many miles per 1 hour?)

14) A part time instructor earned $\$293\frac{3}{4}$ for $12\frac{1}{2}$ hours of teaching. How much would he make for teaching 10 hours?

15) When a sack of beans is $\frac{3}{4}$ filled it weighs 58 pounds. How much will it weigh when it is completely filled?

16) In the People's Republic of China, 3 out of every 10 people are illiterate. Assuming the population of China is 1,037,588,000 people, how many are illiterate?

17) 18 karat gold is composed of 18 parts pure gold out of 24 parts total metal. If a bracelet of 18 kt. gold weighs 16 ounces, how many ounces are pure gold?

18) A computer can print 10 digits every $\frac{2}{25}$th of a second. How long would it take to print the number "pi" to 10,000 digits?

19) The smallest known insects in the world are the "hairy winged" beetles. They measure $\frac{8}{1000}$ of an inch in length. How many of these beetles could we place in a line and fill up one *foot*?

20) In the 1940s, a French architect named Le Corbusier developed a scale of measures that architects and engineers still use today to design structures that accommodate the dimensions of the human body. As a result of Le Corbusier's experiments, it was found that MOST of the time the ratio of the *upper part* of a person's body (from top of the head to the navel) to the *lower part* (from the navel to the bottom of the feet) is 5 to 8 (5:8)

a) Suppose a woman measures 25 inches from her bellybutton to the top of her head. Determine the measurement from her navel to her feet.

b) Measure your own body and check how close you come to Le Corbusier's ratio.

21) If a stack of 1000 \$1. bills measures $\frac{1}{4}$ inch, how high would a stack of 1 trillion dollar bills measure? (Recall 1 trillion = 1,000,000,000,000). The U.S. government budget for 1987 is \$1 trillion.

COMPLEX FRACTIONS

If we think back to our work with mixed numbers, we recall that fractions can be thought of as representing division problems. In fact, every fraction gives us a division problem. Right now we may not know how to get the answers for every problem, but theoretically it can be done.

For example: $\frac{8}{4} = 8 \div 4 \qquad \frac{5}{3} = 5 \div 3$

Similarly, any division problem can be expressed using fraction notation. For example, the division problem $12 \div 3$ can be expressed as $\frac{12}{3}$. If we think of the fraction bar (—) as indicating division, then we can express a problem like $\frac{2}{3} \div \frac{4}{9}$ as a fraction.

It would look like this: $\dfrac{\frac{2}{3}}{\frac{4}{9}}$

> Whenever we have other than whole numbers in the numerator and denominator of a fraction, we have what is called a COMPLEX FRACTION.

Complex fractions usually have a fraction in the numerator and/or denominator.

Our complex fraction above can be simplified by thinking of it as the division problem, where the *denominator* is what gets *inverted*.

$$\text{Therefore}\quad \dfrac{\frac{2}{3}}{\frac{4}{9}} = \frac{2}{3} \div \frac{4}{9} = \frac{2}{3} \times \frac{9}{4} = \frac{\overset{1}{\cancel{2}}}{\underset{1}{\cancel{3}}} \times \frac{\overset{3}{\cancel{9}}}{\underset{2}{\cancel{4}}} = \frac{3}{2}$$

Some complex fractions involve addition or subtraction problems in either numerator or denominator or both.

Let's examine such a problem.

Example 1: $\dfrac{\frac{1}{2}+\frac{1}{4}}{\frac{9}{8}}$

The key to working this problem is to put all expressions in either numerator and denominator as single fractions before division takes place; then the denominator is the divisor and it gets inverted.

First we add $\frac{1}{2}+\frac{1}{4}$ $\quad(\frac{1}{2}+\frac{1}{4}=\frac{2}{4}+\frac{1}{4}=\frac{3}{4})$

Then $\dfrac{\frac{1}{2}+\frac{1}{4}}{\frac{9}{8}} = \dfrac{\frac{3}{4}}{\frac{9}{8}}$ which is $\frac{3}{4}\div\frac{9}{8}$.

$$\dfrac{\frac{3}{4}}{\frac{9}{8}} = \frac{3}{4}\div\frac{9}{8} = \frac{3}{4}\times\frac{8}{9} = \frac{\overset{1}{\cancel{3}}}{\underset{1}{\cancel{4}}}\times\frac{\overset{2}{\cancel{8}}}{\underset{3}{\cancel{9}}} = \frac{1\times 2}{1\times 3} = \frac{2}{3}$$

The answer is $\frac{2}{3}$.

There is an alternative method to working the problem we just completed. It is a bit tricky to understand at first, but keep reviewing the steps until it makes sense.

Multiply all parts of the numerator and denominator of the complex fraction by the *Least Common Multiple* of the denominators.

Since we are doing what is done when *building fractions* (i.e., multiplying top and bottom by the same number), we are *NOT* changing the value of the fraction.

Example 2: Work $\dfrac{\frac{1}{2}+\frac{1}{4}}{\frac{9}{8}}$ the alternative way.

Determine the L.C.M. of all the denominators, 2, 4, and 8.

The L.C.M. is 8.

We now proceed to multiply each fraction by the whole number 8 (or $\frac{8}{1}$).

$$\frac{\frac{1}{2}+\frac{1}{4}}{\frac{9}{8}} \quad \text{becomes} \quad \frac{(8)\frac{1}{2}+(8)\frac{1}{4}}{(8)\frac{9}{8}}$$

Now we cancel each fraction

$$\frac{(8)\frac{1}{2}+(8)\frac{1}{4}}{(8)\frac{9}{8}} = \frac{{}^{4}(\cancel{8})\frac{1}{\cancel{2}_{1}}+{}^{2}(\cancel{8})\frac{1}{\cancel{4}_{1}}}{{}_{1}(\cancel{8})\frac{9}{\cancel{8}_{1}}} = \frac{4+2}{9} = \frac{6}{9}$$

↓

We get whole numbers as numerator and denominator.

Our last step is to reduce our answer $\frac{6}{9}$

$$\frac{6}{9} = \frac{2}{3}$$

Our answer is $\frac{2}{3}$.

Example 3: Work $\dfrac{\frac{1}{3}+\frac{3}{4}}{\frac{7}{8}-\frac{2}{3}}$ the alternative way.

First, we find the L.C.M. of 3, 4 and 8. That L.C.M. is 24.

Next, we proceed to multiply each fraction by the L.C.M. as $\frac{24}{1}$.

$$\frac{\left(\frac{24}{1}\right)\frac{1}{3}+\left(\frac{24}{1}\right)\frac{3}{4}}{\left(\frac{24}{1}\right)\frac{7}{8}-\left(\frac{24}{1}\right)\frac{2}{3}} = \frac{\left(\frac{{}^{8}\cancel{24}}{1}\right)\frac{1}{\cancel{3}_{1}}+\left(\frac{{}^{6}\cancel{24}}{1}\right)\frac{3}{\cancel{4}_{1}}}{\left(\frac{{}^{3}\cancel{24}}{1}\right)\frac{7}{\cancel{8}_{1}}-\left(\frac{{}^{8}\cancel{24}}{1}\right)\frac{2}{\cancel{3}_{1}}} =$$

$$\frac{8(1)+6(3)}{3(7)-8(2)} = \frac{8+18}{21-16} = \frac{26}{5}$$

Our answer is $\frac{26}{5}$ or $5\frac{1}{5}$.

– WORKSHEET –

COMPLEX FRACTIONS

Practice Set A

Simplify the following complex fractions:

1) $\dfrac{\frac{4}{5}}{\frac{27}{10}}$

2) $\dfrac{8}{\frac{2}{5}}$

3) $\dfrac{\frac{9}{16}}{3}$

4) $\dfrac{1\frac{2}{3}}{6\frac{2}{3}}$

5) $\dfrac{3\frac{1}{7}}{1\frac{4}{7}}$

6) $\dfrac{\frac{1}{2} + \frac{1}{4}}{\frac{1}{4} + \frac{1}{8}}$

7) $\dfrac{\frac{2}{3} + \frac{3}{5}}{\frac{4}{15}}$

8) $\dfrac{2 + \frac{1}{3}}{4 - \frac{1}{2}}$

9) $\dfrac{6}{\frac{1}{2} + \frac{2}{3}}$

10) $\dfrac{5 - \frac{1}{2}}{7 + \frac{1}{2}}$

The following complex fractions are a bit more challenging. They have several layers. These type are often called "continued fractions". Simplify them.

11) $\dfrac{1}{\dfrac{1}{2 + \frac{1}{2}}}$

12) $1 + \dfrac{1}{1 + 1}$

13) $1 + \dfrac{1}{1 + \dfrac{1}{1 + 1}}$

14) $\dfrac{1}{2 + \frac{1}{3}}$

15) $\dfrac{1}{2 + \dfrac{1}{3 + \frac{1}{4}}}$

COMPLEX FRACTIONS

Practice Set B

Simplify the following complex fractions:

1) $\dfrac{\frac{4}{3}}{\frac{10}{27}}$

2) $\dfrac{18}{\frac{3}{4}}$

3) $\dfrac{\frac{7}{8}}{2}$

4) $\dfrac{3\frac{3}{4}}{8\frac{1}{4}}$

5) $\dfrac{4\frac{9}{10}}{2\frac{1}{10}}$

6) $\dfrac{\frac{2}{3}+\frac{1}{2}}{\frac{5}{6}-\frac{1}{2}}$

7) $\dfrac{\frac{5}{6}+\frac{4}{5}}{\frac{7}{30}}$

8) $\dfrac{12+\frac{1}{2}}{10-2\frac{1}{2}}$

9) $\dfrac{26}{\frac{2}{3}+\frac{3}{2}}$

10) $\dfrac{9-\frac{3}{4}}{6-\frac{3}{4}}$

The following complex fractions are a bit more complicated. They are called "continued fractions". Simplify them.

11) $\dfrac{1}{2+\frac{1}{2}}$

12) $\dfrac{1}{2+\dfrac{1}{2+\frac{1}{2}}}$

13) $2-\dfrac{1}{1+2}$

14) $2-\dfrac{1}{1+\frac{2}{3}}$

15) $2-\dfrac{2}{2-\frac{2}{3}}$

FRACTIONS AND THE CALCULATOR

Not all calculators have the same logic. I have outlined a method that works on all calculators (except those with Reverse Polish notation). The method shown is not always the simplest, but it does give correct answers on all calculators without reverse Polish operation.

You can work any problem with fractions on the calculator. The only drawback is that calculators do not give fractional answers. Instead, the answers are given in decimal form. There are several common fractions that can be easily recognized, and you would do well to memorize them.

Here are some decimal forms to memorize:

$\frac{1}{4} = .25$

$\frac{1}{2} = .5$

$\frac{3}{4} = .75$

$\frac{1}{3} = .333333$

$\frac{2}{3} = .666666$

$\frac{1}{6} = .166666$

$\frac{5}{6} = .8333333$

$\frac{1}{8} = .125$

$\frac{3}{8} = .375$

$\frac{5}{8} = .625$

$\frac{7}{8} = .875$

$\frac{1}{5} = .2$

$\frac{2}{5} = .4$

$\frac{3}{5} = .6$

$\frac{4}{5} = .8$

$\frac{1}{10} = .1$

$\frac{3}{10} = .3$

$\frac{7}{10} = .7$

$\frac{9}{10} = .9$

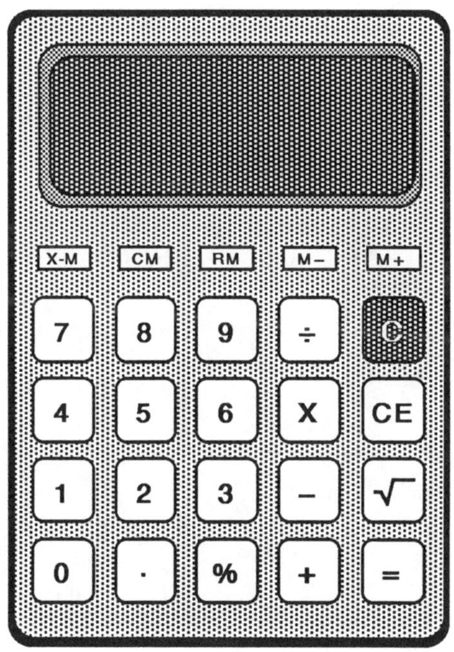

On the calculator we will use these buttons:

(+)	(−)	(×)	(÷)	(=)	(C)	(STO) or (M+)*	(RCL) or (MR)
add	subtract	multiply	divide	equals	clear	store	recall

* (M+) is usually the button for cumulative memory.

Before we can compute with fractions on the calculator, they must be changed to decimals.

I. Changing Fractions to Decimals

You can put a fraction into decimal form if you consider it as a division expression. You enter the fraction as numerator *divided by* denominator. (numerator ÷ denominator)

Example 1: Enter the fraction $\frac{3}{4}$

We type 3 (÷) 4 (=) The answer appears in the display as .75

Example 2: Change $2\frac{3}{5}$ to decimal form.

Since $2\frac{3}{5}$ is a mixed number, we will enter the fraction part first. (Some calculators have algebraic order and the problem can be entered just as it reads.)

We type 3 (÷) 5 (+) 2 (=)

The display will read 2.6

After doing any operation on the calculator, it is a safeguard to always press the (=) button before continuing.

II. Multiplying Fractions

To multiply fractions on the calculator, we need only keep track of which numbers are numerators, and which numbers are denominators. We multiply the numerators first, then we divide by EACH denominator in the problem.

Example 3: Work $\frac{1}{2} \times \frac{1}{4}$ on the calculator.

We press the following buttons:

1 (×) 1 (÷) 2 (÷) 4 (=)

The answer on the calculator display will be .125 (which we memorized as $\frac{1}{8}$.

There is another sequence we can follow. We need only remember to multiply by *any* value in a numerator and divide by *any* value in a denominator.

Here is another sequence for the same problem:

1 (÷) 2 (×) 1 (÷) 4 (=)

Example 4: Multiply $2\frac{3}{5} \times \frac{1}{2}$ on the calculator.

Since there is a mixed number in the problem, we need to work that first. Then we multiply by the fraction.

We type the following:

3 (÷) 5 (+) 2 (=) (×) 1 (÷) 2 (=)

The answer will appear in the display as 1.3 which is $1\frac{3}{10}$.

Example 5: Multiply $2\frac{3}{5} \times 1\frac{1}{2}$

Since there are two mixed numbers in this problem, we will need to store one of them in the memory of the calculator. We will recall it after the other mixed number is entered. Most calculators have a STORE button

(either (STO) or (M+)

The recall button is labelled either (RCL) or (MR)).

Here is the sequence of steps:

3 (÷) 5 (+) 2 (=) (STO) then

1 (÷) 2 (+) 1 (=) (×) (RCL) (=)

The answer will appear in the display as 3.9 (which is $3\frac{9}{10}$).

Example 6: Multiply $3\frac{1}{3}$ by $4\frac{1}{2}$

Remember to first store one of the mixed numbers.
Here is the sequence of steps:

1 (÷) 3 (+) 3 (=) (STO) then

1 (÷) 2 (+) 4 (=) (×) (RCL) (=)

The answer will appear in the display as 15. (Some calculators show the answer as 14.9999. This is an idiosyncracy of the logic of that particular calculator. It means 15.)

III. Dividing Fractions

Dividing fractions on the calculator is almost the same as multiplying fractions. It helps to rewrite a division problem as a multiplication problem before entering it into the calculator.

Example 7: Divide $\frac{3}{4} \div \frac{9}{4}$ on the calculator.

If we rewrite the problem as $\frac{3}{4} \times \frac{4}{9}$ first, then we can enter it in the calculator as a multiplication problem.

We type: 3 (×) 4 (÷) 4 (÷) 9 (=) or

3 (÷) 4 (×) 4 (÷) 9 (=)

The answer appears in the display as .33333333 (the number of 3s depends on the calculator). This is the fraction $\frac{1}{3}$

Another way we can do the problem directly on the calculator is if we remember to switch the numerator and denominator of the $\frac{9}{4}$ as they are put into the calculator.

We would type 3 (÷) 4 (×) 4 (÷) 9 (=)

(9 ↑ new denominator; 4 ↓ new numerator)

The 9 becomes the NEW denominator (hence it appears after the (÷) button.)

The 4 becomes the NEW numerator (hence it appears after the (×) button.)

Example 8: Divide $7\frac{1}{2} \div 1\frac{1}{2}$

We must store the mixed number that appears *after* the ÷ sign.

The order of entry into the calculator is important. We store into memory the mixed number that appears AFTER the ÷ sign. We enter the other mixed number, then divide by the RECALL number. Here are the steps:

1 (÷) 2 (+) 1 (=) (STO)

1 (÷) 2 (+) 7 (÷) (RCL) (=)

The answer will appear in the display as 5.

IV. Adding Fractions

In order to add fractions we will have to use the STORE (STO) capability of the calculator. It is easiest to add two fractions at a time. We usually enter one of the fractions into memory (using the STO button), enter the other fraction into the calculator, and then add the stored fraction by pressing the RCL button.

Example 9: Add $\frac{1}{2} + \frac{1}{4}$

We know the answer is $\frac{3}{4}$ or .75

These are the steps on the calculator:

1 (÷) 2 (=) (STO)

1 (÷) 4 (=) (+) (RCL) (=)

The display will have .75 as the answer.

Example 10: Add $\frac{1}{3} + \frac{1}{2} + \frac{2}{5}$

Let's first add the $\frac{1}{3}$ and the $\frac{1}{2}$. Here are the steps:

1 (÷) 3 (=) (STO)

1 (÷) 2 (=) (+) (RCL) (=) (STO)

We add the result of having added the first two fractions. Then we proceed to add the third fraction.

We do: 2 (÷) 5 (=) (+) (RCL) (=)

Our final answer will appear in the display as 1.2333333 (all threes at the end). This is the decimal form of the fraction $\frac{37}{30}$.

Some calculators have a cumulative memory so this problem can be done by entering each fraction and continually putting the result into *cumulative* additive memory. Usually the memory button is indicated as (M+) . Do *not* work the problem as above if you have *cumulative* memory. Just write down the first addition result, then add that to the third. Here is how we could work the above problem using that procedure for *cumulative* memory:

1 (÷) 3 (=) (M+)

1 (÷) 2 (=) (M+)

2 (÷) 5 (+) (M+) (RCL)

The display should show the same answer: 1.2333333

Example 11: Add $2\frac{3}{4} + 1\frac{1}{2}$

We enter the first mixed number into the calculator memory. Then we add the second mixed number TO that recalled (RCL) first number. These are the steps:

3 (÷) 4 (+) (2) (=) (STO)

1 (÷) 2 (+) 1 (=) (+) (RCL) (=)

The answer is 4.25 or $4\frac{1}{4}$.

V. Subtracting Fractions

When subtracting fractions it is important which fraction is entered AFTER the subtraction sign (−). Every calculator has its own shortcuts, but the method I am outlining will work with most calculators. ALWAYS ENTER THE FRACTION *AFTER* THE SUBTRACTION SIGN (−), THEN STORE IT.

Example 12: Subtract: $\frac{1}{2} - \frac{1}{4}$

ENTER THE $\frac{1}{4}$ FIRST, THEN *STORE* IT. Then you enter the first fraction and subtract the RCL number. Here are the steps:

1 (÷) 4 (=) (STO)

1 (÷) 2 (=) (−) (RCL) (=)

The answer is .25 (or $\frac{1}{4}$).

Example 13: Subtract $2\frac{3}{4} - 1\frac{7}{8}$

Enter the $1\frac{7}{8}$ *first* (since it appears AFTER the − sign). Store it.

Then enter the $2\frac{3}{4}$.

Subtract the RCL fraction which was stored ($1\frac{7}{8}$).

Here are the steps:

7 (÷) 8 (+) 1 (=) (STO)

3 (÷) 4 (+) 2 (=) (−) (RCL) (=)

The answer in the display is .875.

Example 14: $14\frac{1}{2} - 3\frac{1}{4} - 2\frac{1}{8}$

This problem should be done two fractions at a time. You will need to STORE the result of subtracting the first two fractions (or write down the result).

Here are the steps:

1 (÷) 4 (+) 3 (=) (STO)

1 (÷) 2 (+) 14 (=) (−) (RCL) (=) (STO)

1 (÷) 8 (+) 2 (=) (±)* (+) (RCL) (=)

*The (±) button puts the subtraction sign in front of the $2\frac{1}{8}$.

The answer appears in the display as 9.125 which is $9\frac{1}{8}$.

(Do not use this method if you have *cumulative* memory. You will have to work the first 2 fractions, write down the result, then work the next part, i.e.:

$14\frac{1}{2} - 3\frac{1}{4}$ = first result

then (first result) − $2\frac{1}{8}$)

IMPORTANT RULES FOR USING THE CALCULATOR TO DO FRACTION PROBLEMS:

1) Fractions are always entered as (numerator) ÷ (denominator).

Always press the (=) button as a safeguard after entering the fraction.

Example: $\frac{3}{4}$ is entered as: 3 (÷) 4 (=)

2) When entering MIXED NUMBERS, enter the fraction part first, then ADD the whole number part.

Example: $2\frac{5}{6}$ is entered as: 5 (÷) 6 (+) 2 (=)

> NOTE: Many calculators use algebraic order of operations so the mixed number can be directly entered as 2 + 5 ÷ 6 =

3) When doing multiplication problems, any numerator is entered AFTER the (×) button, and any denominator is entered AFTER the (÷) button.

Example: $\frac{3}{8} \times \frac{4}{9}$ is done as: 3 (÷) 8 (×) 4 (÷) 9 (=)

8 ↓ denom.; 4 ↓ num.; 9 ↓ denom.

4) Memorize the common decimal forms of fractions. You can then change calculator answers to correct fraction form.

Example: The calculator reads out 3.375

We know that it is the mixed number $3\frac{3}{8}$.

FRACTIONS ON THE CALCULATOR

Practice Set A

Use your calculator to get the decimal form of the following fractions and/or mixed numbers:

1) $\frac{3}{40}$
2) $\frac{8}{3}$
3) $\frac{1}{12}$
4) $\frac{5}{12}$
5) $\frac{19}{5}$
6) $\frac{11}{20}$
7) $\frac{17}{30}$
8) $\frac{57}{8}$
9) $\frac{11}{12}$
10) $\frac{100}{3}$
11) $7\frac{5}{8}$
12) $\frac{30}{55}$
13) $4\frac{5}{9}$
14) $19\frac{3}{4}$
15) $89\frac{5}{11}$
16) $66\frac{2}{3}$
17) $\frac{2345}{100000}$
18) $12\frac{1}{2}$
19) $\frac{1}{7}$
20) $\frac{2}{7}$
21) $\frac{3}{7}$
22) $\frac{4}{7}$
23) $\frac{5}{7}$
24) $\frac{6}{7}$
25) $\frac{99}{100}$

Multiply or Divide as the problem asks using your calculator. Check your answers by doing the problem long hand:

26) $1000 \times \frac{10}{3}$
27) $\frac{1}{2} \times \frac{2}{3} \times \frac{3}{4}$
28) $\frac{1}{5} \div \frac{1}{5} \div \frac{1}{5} \div \frac{1}{5}$
29) $2\frac{3}{8} \times 1\frac{1}{2}$
30) $\frac{11}{100} \div \frac{2}{5}$
31) $8\frac{3}{4} \div 1\frac{3}{4}$
32) $\frac{7}{16} \times \frac{4}{3} \times 9$
33) $\frac{1}{2} \div \frac{3}{4}$
34) $\frac{5}{8} \times \frac{4}{5} \div \frac{9}{10}$
35) $11\frac{1}{2} \div 2\frac{3}{4}$

36) $3\frac{1}{8} \times 2\frac{3}{8} \times 5\frac{1}{4}$

37) $8\frac{1}{2} \div 4\frac{1}{4} \div \frac{51}{102}$

38) $\frac{17}{3} \times \frac{3}{2} \div \frac{5}{2}$

39) $1\frac{3}{4} \times \frac{2}{3}$

40) $2\frac{3}{5} \div \frac{6}{5}$

41) $(\frac{1}{2})^3$

42) $(1\frac{2}{3})^2$

Add or subtract the following by using your calculator. Check your answers by working the exercises out long hand.

43) $\frac{1}{2} + \frac{3}{4}$

44) $\frac{1}{2} + \frac{2}{3}$

45) $4\frac{1}{5} + 2\frac{7}{8}$

46) $\frac{4}{5} - \frac{2}{3}$

47) $\frac{4}{5} - \frac{2}{3} - \frac{1}{30}$

48) $8 - \frac{5}{8}$

49) $\frac{1}{3} + 12\frac{2}{3}$

50) $11\frac{7}{32} - 9\frac{3}{32}$

51) $5\frac{3}{10} + 4\frac{7}{10}$

52) $99\frac{5}{6} - 85\frac{5}{12}$

53) $1 - \frac{1}{2} - \frac{1}{4} - \frac{1}{8}$

54) $10 - 9\frac{7}{9} + \frac{2}{9}$

55) $10 - 9\frac{7}{9} - \frac{2}{9}$

56) $10 - (9\frac{7}{9} + \frac{2}{9})$

57) $7\frac{3}{8} - (2\frac{3}{4} - 1\frac{1}{2})$

Verify that the following sentences are true by using the calculator:

58) $1\frac{1}{2} \times 3 = 1\frac{1}{2} + 3$

59) $1\frac{1}{3} \times 4 = 1\frac{1}{3} + 4$

60) $1\frac{1}{4} \times 5 = 1\frac{1}{4} + 5$

61) $1\frac{1}{5} \times 6 = 1\frac{1}{5} + 6$

62) $1 \times \frac{1}{2} = 1 - \frac{1}{2}$

63) $2 \times \frac{2}{3} = 2 - \frac{2}{3}$

64) $3 \times \frac{3}{4} = 3 - \frac{3}{4}$

65) $4\frac{1}{2} \div 3 = 4\frac{1}{2} - 3$

66) $5\frac{1}{3} \div 4 = 5\frac{1}{3} - 4$

67) $6\frac{1}{4} \div 5 = 6\frac{1}{4} - 5$

FRACTIONS ON THE CALCULATOR

Practice Set B

Use your calculator to get the decimal form of the following fractions and/or mixed numbers:

1) $\frac{3}{50}$ 2) $\frac{7}{3}$ 3) $\frac{1}{16}$ 4) $\frac{7}{12}$

5) $\frac{21}{5}$ 6) $\frac{11}{25}$ 7) $\frac{19}{40}$ 8) $\frac{35}{8}$

9) $\frac{13}{16}$ 10) $\frac{200}{3}$ 11) $8\frac{3}{4}$ 12) $\frac{33}{55}$

13) $4\frac{4}{9}$ 14) $21\frac{3}{8}$ 15) $98\frac{4}{11}$ 16) $33\frac{1}{3}$

17) $12\frac{1}{8}$ 18) $\frac{1234}{10000}$ 19) $\frac{1}{11}$ 20) $\frac{1}{111}$

21) $\frac{2}{11}$ 22) $\frac{2}{111}$ 23) $\frac{2}{101}$ 24) $\frac{2}{1111}$

25) $2\frac{909}{1000}$

Multiply or Divide using your calculator. Check your answers by doing the problem long hand.

26) $200 \times \frac{20}{3}$ 27) $\frac{1}{2} \times \frac{4}{5} \times \frac{5}{6}$

28) $\frac{1}{2} \div \frac{1}{2} \div \frac{1}{2} \times \frac{1}{2}$ 29) $4\frac{2}{3} \times 1\frac{3}{7}$

30) $\frac{99}{100} \div \frac{33}{25}$ 31) $4\frac{2}{5} \div 1\frac{1}{10}$

32) $\frac{5}{12} \times \frac{4}{3} \times 18$ 33) $\frac{2}{3} \div \frac{5}{6}$

34) $\frac{3}{8} \times \frac{4}{9} \div \frac{21}{10}$ 35) $12\frac{1}{2} \div 1\frac{1}{14}$

36) $1\frac{1}{8} \times 1\frac{10}{35} \times 7\frac{7}{9}$

37) $3\frac{1}{2} \div 2\frac{1}{3} \div \frac{34}{85}$

38) $1\frac{7}{8} \times \frac{4}{5}$

39) $(\frac{3}{4})^2$

40) $(1\frac{1}{2})^3$

Add or subtract the following by using your calculator. Check your answers by working out the exercises long hand.

41) $\frac{1}{2} + \frac{4}{5}$

42) $\frac{3}{4} + \frac{2}{3}$

43) $5\frac{1}{10} + 3\frac{2}{5}$

44) $\frac{11}{24} - \frac{5}{18}$

45) $\frac{11}{24} - \frac{5}{18} - \frac{5}{36}$

46) $9 - \frac{5}{9}$

47) $\frac{5}{8} + 15\frac{3}{8}$

48) $21\frac{4}{7} - 9\frac{2}{7}$

49) $55\frac{9}{10} + 44\frac{1}{10}$

50) $10 - 9\frac{3}{11} + \frac{8}{11}$

51) $10 - 9\frac{3}{11} - \frac{8}{11}$

52) $10 - (9\frac{3}{11} + \frac{8}{11})$

53) $15\frac{4}{5} - (12\frac{3}{8} + 2\frac{3}{4})$

FRACTIONS — REVIEW EXERCISES

Practice Set A

Change each of the following to mixed number form. Reduce if possible.

1) $\frac{19}{8}$ 2) $\frac{105}{25}$ 3) $\frac{77}{10}$ 4) $\frac{97}{8}$ 5) $\frac{2001}{999}$

Change each of the following to improper fraction form.

6) $7\frac{1}{2}$ 7) $2\frac{3}{4}$ 8) $16\frac{2}{3}$ 9) $9\frac{9}{10}$ 10) $50\frac{1}{2}$

Build fractions by finding the missing number:

11) $\frac{5}{6} = \frac{?}{18}$ 12) $\frac{7}{8} = \frac{56}{?}$

13) $11\frac{3}{4} = \frac{?}{16}$ 14) $2\frac{2}{3} = \frac{?}{15}$

15) $12\frac{1}{2} = \frac{100}{?}$

Reduce each of the following to lowest terms:

16) $\frac{18}{32}$ 17) $\frac{34}{51}$ 18) $\frac{35}{63}$ 19) $\frac{125}{375}$ 20) $\frac{16}{18}$

21) $\frac{100}{45}$ 22) $\frac{24}{60}$ 23) $\frac{28}{77}$ 24) $\frac{180}{240}$ 25) $\frac{39}{1001}$

Order the following fractions from LARGEST to SMALLEST:

26) $\frac{3}{8}, \frac{5}{12}, \frac{1}{3}$ 27) $\frac{4}{9}, \frac{44}{100}, \frac{43}{99}, \frac{439}{999}$

28) $1\frac{1}{2}, 1\frac{3}{7}, 1\frac{5}{14}, 1\frac{36}{70}$

Add and Subtract as indicated in the following:

29) $\begin{array}{r}\frac{3}{16}\\ +\frac{5}{16}\\ \hline\end{array}$

30) $\begin{array}{r}\frac{7}{10}\\ +\frac{5}{3}\\ \hline\end{array}$

31) $\frac{4}{5} + \frac{2}{3} + \frac{5}{6}$

32) $\frac{7}{9} + \frac{5}{6} - \frac{1}{2}$

33) $5\frac{1}{2} + 3\frac{2}{3}$

34) $\begin{array}{r}7\frac{3}{5}\\ +9\frac{1}{2}\\ \hline\end{array}$

35) $\begin{array}{r}8\frac{5}{7}\\ -6\frac{2}{3}\\ \hline\end{array}$

36) $\begin{array}{r}12\frac{1}{4}\\ -\ 8\frac{1}{3}\\ \hline\end{array}$

37) $10\frac{3}{4} + 1\frac{1}{2} - 7\frac{5}{12}$

Multiply and Divide as indicated in the following:

38) $\frac{3}{5} \times \frac{4}{9} \times \frac{35}{12}$

39) $\frac{3}{8} \div \frac{9}{16}$

40) $4 \div \frac{3}{5}$

41) $\frac{1}{2} \div 7$

42) $6\frac{2}{3} \times 2\frac{1}{4}$

43) $3\frac{2}{3} \div 2\frac{4}{9}$

44) $\frac{2}{3} \div \frac{2}{3} \div \frac{2}{3}$

45) $(\frac{3}{4})^3$

46) $(2\frac{1}{2})^2$

47) $2 \div 4 \div 3$

48) A contractor needs to pour 18 cubic feet of concrete. If one sack of concrete will pour $\frac{2}{3}$ of a cubic foot, how many sacks will be needed?

49) How much larger than $7\frac{5}{6}$ is $7\frac{6}{5}$?

50) Your sexy new sports car went $569\frac{1}{2}$ miles on $16\frac{3}{4}$ gallons of gasoline. How many miles did you get per gallon?

51) One meter is approximately $39\frac{1}{3}$ inches. How many meters does a length of $91\frac{7}{9}$ inches measure?

52) Last month your electricity bill showed that you used $702\frac{2}{3}$ kilowatt-hours during a 62 day period. Assuming that you will use electricity at the same rate this month, how many kilowatt-hours will you use by the end of the next 30 day month?

53) Out of a class of 45 students, $\frac{2}{5}$ received an A grade, $\frac{4}{15}$ received a B grade, and $\frac{2}{15}$ received a C grade. The rest of the class received a D.

a) What fraction of the class received a D?

b) How many students received a D?

FRACTIONS — REVIEW EXERCISES

Practice Set B

Change each of the following to mixed number form. Reduce if possible.

1) $\frac{21}{5}$ 2) $\frac{85}{15}$ 3) $\frac{27}{8}$ 4) $\frac{100}{30}$ 5) $\frac{136}{12}$

Change each of the following to improper fraction form.

6) $9\frac{2}{3}$ 7) $4\frac{7}{16}$ 8) $24\frac{3}{4}$ 9) $10\frac{7}{8}$ 10) $1001\frac{98}{99}$

Build fractions by finding the missing number:

11) $\frac{3}{4} = \frac{?}{100}$ 12) $\frac{9}{10} = \frac{27}{?}$

13) $1\frac{11}{24} = \frac{?}{72}$ 14) $15\frac{3}{5} = \frac{?}{25}$

15) $2\frac{1}{4} = \frac{72}{?}$

Reduce each of the following to lowest terms:

16) $\frac{150}{200}$ 17) $\frac{55}{121}$ 18) $\frac{12}{15}$ 19) $\frac{32}{36}$ 20) $\frac{90}{25}$

21) $\frac{144}{240}$ 22) $\frac{24}{28}$ 23) $\frac{36}{90}$ 24) $\frac{44}{24}$ 25) $\frac{1001}{52}$

Order the following fractions from LARGEST to SMALLEST:

26) $\frac{4}{9}, \frac{11}{36}, \frac{23}{72}$ 27) $\frac{7}{10}, \frac{5}{6}, \frac{13}{15}, \frac{2}{3}$

28) $1\frac{1}{6}, 1\frac{4}{25}, 1\frac{7}{30}, 1\frac{13}{100}$

Add and Subtract as indicated:

29) $\begin{array}{r} \frac{2}{15} \\ +\frac{7}{15} \\ \hline \end{array}$

30) $\begin{array}{r} \frac{7}{18} \\ +\frac{11}{12} \\ \hline \end{array}$

31) $\frac{5}{12} + \frac{2}{3} + \frac{4}{9}$

32) $\frac{6}{5} - \frac{2}{3} + \frac{2}{9}$

33) $6\frac{3}{4} + 2\frac{2}{9}$

34) $\begin{array}{r} 15\frac{2}{3} \\ +\ 9\frac{3}{5} \\ \hline \end{array}$

35) $\begin{array}{r} 9\frac{8}{9} \\ -7\frac{7}{8} \\ \hline \end{array}$

36) $\begin{array}{r} 22\frac{2}{5} \\ -19\frac{3}{4} \\ \hline \end{array}$

37) $2\frac{1}{2} + 3\frac{2}{3} - 5\frac{7}{12}$

Multiply and Divide as indicated in the following:

38) $\frac{4}{15} \times \frac{3}{2} \times \frac{35}{18}$

39) $3\frac{1}{5} \div \frac{8}{25}$

40) $12 \div \frac{5}{6}$

41) $\frac{2}{3} \div 6$

42) $5\frac{1}{3} \times 8\frac{1}{4}$

43) $5\frac{1}{10} \div 1\frac{2}{15}$

44) $\frac{1}{4} \div \frac{3}{4} \div \frac{1}{4}$

45) $(\frac{1}{10})^3$

46) $(2\frac{2}{3})^2$

47) $12 \div 30 \div 3$

48) How much larger is $2\frac{5}{7}$ than $2\frac{2}{3}$?

49) In a rectangle, the length is $1\frac{3}{4}$ inches longer than the width. If the width is $8\frac{2}{3}$ inches long, find the perimeter of the rectangle (add up all the side lengths).

50) If you exchange $21 U.S. dollars for 15 British pounds, what is the exchange rate per dollar?

51) The largest lobster ever caught weighed $44\frac{3}{8}$ pounds. It was sold to a restaurant in Bayville, New York. Assuming that was all meat, how many "lobster specials" could have come from that crustacean if we know that the special serves $1\frac{1}{4}$ pounds of lobster meat?

52) If the U.S. population is approximately $\frac{1}{20}$ of the total world population, and $\frac{3}{4}$ of the U.S. population live in cities, assuming that the world population is 4,864 million people,

a) What is the population of the United States?
b) How many people in the United States live in cities?

FRACTIONS — REVIEW OF IMPORTANT WORDS

Here is a checklist of words that are used in dealing with fractions. See if you can identify each one and define them to yourself.

numerator

denominator

unit fraction

proper fraction

improper fraction

mixed number

least common denominator

greatest common factor

reducing to lowest terms

building fractions

equivalent fractions

cross-wise reducing

reciprocal

complex fraction

Remember: ALL ANSWERS TO FRACTION PROBLEMS SHOULD BE REDUCED TO LOWEST TERMS.

SAMPLE EXAMINATION #1 FOR FRACTIONS

This exam would be worth 100 points and would be given during a 50 minute period.

1) Reduce:

(a) $\frac{20}{48} = \_\_\_\_$

(b) $\frac{125}{500} = \_\_\_\_$

2) $8\frac{2}{3} + 3\frac{1}{3}$

3) $12\frac{3}{4} + 10\frac{1}{2}$

4) $3\frac{1}{2} + 2\frac{3}{4} + 7\frac{1}{4}$

5) $4\frac{2}{3} + 6\frac{5}{8}$

6) $4\frac{5}{6} + 3\frac{3}{8} + 7\frac{2}{3}$

7) $23\frac{7}{8} - 18\frac{1}{2}$

8) $9\frac{7}{8} + 4\frac{3}{4}$

9) $11\frac{1}{3} - 7\frac{3}{4}$

10) $12\frac{1}{2} - 8\frac{2}{3}$

11) $\frac{2}{3} \times \frac{4}{5}$

12) $\frac{7}{8} \times \frac{4}{5}$

13) $2\frac{1}{2} \times \frac{7}{15}$

14) Find $\frac{5}{6}$ of 216.

15) $\frac{7}{8} \div \frac{1}{2}$

16) $\frac{1}{3} \div 1\frac{1}{2}$

17) $4\frac{1}{2} \div 1\frac{2}{3}$

18) $2\frac{1}{4} \div 2\frac{4}{7}$

19) $\frac{1}{2} \times \frac{2}{3} \times \frac{3}{4}$

20) $12 \times 1\frac{1}{2} \times \frac{3}{4}$

21) Jim worked $8\frac{1}{2}$ hours on Monday, $6\frac{3}{4}$ hours Tuesday, $7\frac{1}{2}$ hours Wednesday, 6 hours Thursday, and $8\frac{2}{3}$ hours Friday. If he was supposed to have worked 40 hours, how many is he short?

22) Joan drove her car at $40\frac{1}{2}$ miles per hour for $3\frac{3}{4}$ hours. Find how far she travelled during that time.

23) Find the distance around this field.

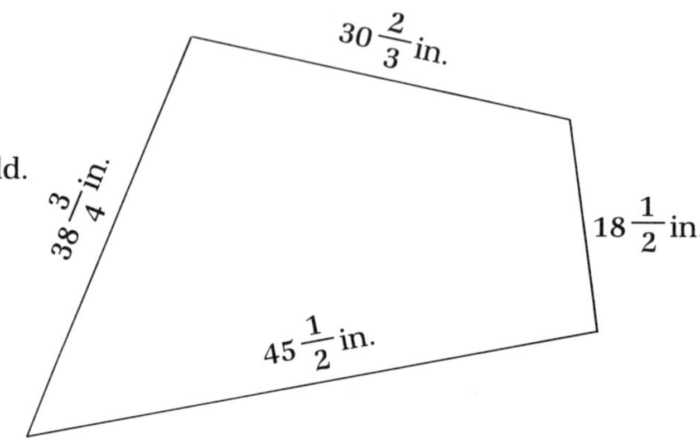

24) How many boards of length $2\frac{3}{4}$ ft. can be cut from a board which is 22 ft. long? Disregard waste.

25) A man bought $12\frac{3}{4}$ lbs. of nails at $24\frac{1}{2}$¢ per pound. Find his bill. Leave your answer as a mixed number of cents.

SAMPLE EXAMINATION #2
FOR FRACTIONS

Please show all your work on this paper. REDUCE all answers.

1) Reduce:

(a) $\frac{16}{30}$ = _____

(b) $\frac{375}{1000}$ = _____

2) $6\frac{2}{5} + 3\frac{4}{5}$

3) $9\frac{3}{4} + 3\frac{1}{8}$

4) $6\frac{1}{3} + 2\frac{1}{2} + 3\frac{3}{4}$

5) $4\frac{2}{3} + 8\frac{3}{8}$

6) $4\frac{5}{6} + 3\frac{3}{4} + 2\frac{7}{8}$

7) $15\frac{7}{8} - 12\frac{3}{8}$

8) $2\frac{2}{3} - 1\frac{3}{4}$

9) $8\frac{1}{5} - 5\frac{1}{4}$

10) $16 - 10\frac{7}{8}$

11) $\frac{2}{5} \times \frac{7}{10}$

12) $\frac{7}{8} \times \frac{4}{6}$

13) $3\frac{2}{3} \times \frac{1}{2}$

14) Find $\frac{3}{4}$ of 220.

15) $\frac{3}{4} \div \frac{1}{2}$

16) $1\frac{1}{2} \div 2\frac{2}{3}$

17) $2\frac{2}{5} \div 1\frac{1}{5}$

18) $\frac{1}{3} \times \frac{3}{4} \times \frac{5}{6}$

19) $15 \times 1\frac{1}{2} \times \frac{2}{3}$

20) $7\frac{1}{3} \div 2\frac{2}{9}$

21) A box weighing $10\frac{2}{3}$ ounces contains bottles of perfume, each weighing $1\frac{1}{3}$ ounces. How many bottles are in the box?

22) Suppose $\frac{1}{3}$ of a sack of sugar is used for candy, and $\frac{2}{5}$ of a sack of sugar is used for pies. What part of the sack of sugar is left?

23) Jim rides his bicycle $3\frac{1}{4}$ miles, then walks the rest of the way to school. Jim does the same on his return home. If the total distance traveled back and forth to school is $8\frac{2}{3}$ miles, how far does he *walk* each way?

24) How much gasoline can you buy in Mexico for 100 pesos, if $3\frac{1}{3}$ gallons cost you 30 pesos?

SAMPLE EXAMINATION #3
FOR FRACTIONS

The answers to this exam are NOT in back.
Show all work for partial credit. Double check your answer.
Put all answers in simplified (reduced) form, or mixed number form.

1) Reduce:

(a) $\frac{34}{51} =$ ____

(b) $\frac{20}{75} =$ ____

2) $6\frac{1}{4} + 11\frac{3}{4}$

3) $13\frac{2}{3} + 6\frac{2}{3}$

4) $17\frac{1}{2} + 23\frac{3}{4}$

5) $99\frac{7}{12} + \frac{8}{12}$

6) $4\frac{7}{8} + 3\frac{1}{4} + 8\frac{2}{3}$

7) $25\frac{7}{8} - 21\frac{3}{4}$

8) $9 - 2\frac{5}{12}$

9) $6\frac{1}{5} - 5\frac{1}{3}$

10) $13\frac{1}{2} - 8\frac{7}{8}$

11) $\frac{5}{8} \times \frac{2}{3} =$

12) $\frac{1}{2} \times \frac{1}{2} \times \frac{1}{2} =$

13) $8\frac{3}{4} \times \frac{5}{7}$

14) Find $\frac{2}{3}$ of 66.

15) $\frac{2}{5} \div 1\frac{1}{10}$

16) $1\frac{1}{2} \div 3$

17) $5\frac{2}{6} \div \frac{4}{18}$

18) $\frac{8}{9} \times 18$

19) $2 \times \frac{3}{4} \div \frac{7}{8}$

20) Order each of the following fractions from *smallest* to largest. List them in that order.

$\frac{3}{4}, \frac{7}{8}, \frac{17}{16}, \frac{19}{24}$

21) An automobile uses $11\frac{3}{4}$ gallons to go 235 miles. How many miles per gallon did it go?

23) A solution contains $3\frac{3}{8}$ pounds of salt per gallon of water. How many pounds of salt would be used with 225 gallons of water?

24) The length of a rectangular field is $142\frac{1}{2}$ yards. The width of the field is $72\frac{3}{4}$ yards.

a) The field is how much longer than it is wide?

b) How many feet of fence material is needed to enclose the field?

"HOLD ON, LET ME TRY AGAIN....I KNOW IT'S NOT $.78 AND I KNOW IT'S NOT 487.62........."

CHAPTER 3: DECIMALS

READING DECIMALS, DECIMALS TO EXPANDED FORM, AND DECIMALS TO FRACTIONS EXERCISES:

When we read the names of whole numbers, we think of each numeral as naming a specific location which is a power of 10. All the names on the place value chart at the beginning of the workbook are numbers *greater* than 1 or equal to 1 (such as ONES, TENS, HUNDREDS, etc.)

In our base-10 system we also use numerals to hold the place of values that are *fractional, less than 1 whole,* and also based on 10s.

We indicate that we have a fractional part by placing a decimal point, after which comes the fractional values.

Looking at the chart, you see the names of these place values.

	tenths	hundredths	thousandths	10-thousandths	100-thousandths	millionths	10-millionths	100-millionths	billionths	etc.	etc.
• ↑ decimal point											

Here are the fractions that go along with each name:

tenth	$\frac{1}{10}$
hundredth	$\frac{1}{100}$
thousandth	$\frac{1}{1000}$
ten-thousandth	$\frac{1}{10{,}000}$
hundred-thousandth	$\frac{1}{100{,}000}$
millionth	$\frac{1}{1{,}000{,}000}$

etc.

Numbers after the decimal point are read as if there were no point there, but we attach the rightmost place value to the number.

Example 1: .345 is read as 345 *thousandths* since the 5 holds the THOUSANDTHS place.

.04 is read as 4 *hundredths* since 4 is in the second place.

.1001 is read as 1001 *ten-thousandths* since the last 1 holds the ten-thousandths place.

123.40065 is read as 123 and 40,065 hundred-thousandths.

The decimal point is always read as "and".

We can look at a decimal number and get an *expanded form* of the number by using the place value of each numeral.

For example, in the decimal .645, the 6 is in the *tenths* place; the 4 is in the *hundredths* place and the 5 is in the *thousandths* place. The expanded form would tell us we have:

6 tenths plus 4 hundredths plus 5 thousandths.

This can be written as $\underbrace{\frac{6}{10} + \frac{4}{100} + \frac{5}{1000}}_{\text{expanded form}}$.

Example 2: Give the expanded form of .703

$$.703 = \frac{7}{10} + \frac{0}{100} + \frac{3}{1000}$$

Example 3: Give the expanded form of 45.79

$$45.79 = 4(10) + 5(1) + \frac{7}{10} + \frac{9}{100}$$

$$\begin{array}{ccccccc} & \downarrow & \downarrow & \downarrow & \downarrow & \downarrow \\ \text{(Note:} & 4 & 5 & . & 7 & 9) \end{array}$$

Example 4: Give the expanded form of 100.001

$$100.001 = 1(100) + 0(10) + 0(1) + \frac{0}{10} + \frac{0}{100} + \frac{1}{1000}$$

$$\begin{array}{cccccccc} & \downarrow & \downarrow & \downarrow & \downarrow & \downarrow & \downarrow & \downarrow \\ \text{(NOTE:} & 1 & 0 & 0 & . & 0 & 0 & 1) \end{array}$$

We can *change decimals to fractions* easily by reading the decimal number and writing it as it reads.

Example 5: .409 reads as 409-thousandths $= \frac{409}{1000}$

2.008 reads as 2 *and* 8-thousandths $= 2 + \frac{8}{1000} = 2\frac{8}{1000}$

(We can reduce the fraction so $2\frac{8}{1000} = 2\frac{1}{125}$ or $\frac{251}{125}$)

675.85 is read as 675 *and* 85-hundredths $= 675 + \frac{85}{100} = 675\frac{85}{100}$

(We can reduce the fraction part so $675\frac{85}{100} = 675\frac{17}{20}$ or $\frac{13517}{20}$

Example 6: Change the decimal .125 to a fraction in reduced form.

.125 is read as 125-thousandths $= \frac{125}{1000} = \frac{1}{8}$ by reducing

since $\frac{125}{1000} = \frac{1}{8}$. (numerator $\div 125$, denominator $\div 125$)

Example 7: Change $2.\overline{3333333}$ to a fraction.

At this point we can do this ONLY because we memorized the decimal form of $\frac{1}{3}$ in the section "Fractions and the Calculator". Later on we will derive the $\frac{1}{3}$.

The fraction is $2 + \frac{1}{3} = 2\frac{1}{3}$ or $\frac{7}{3}$.

READING DECIMALS, DECIMALS TO EXPANDED FORM, AND DECIMALS TO FRACTIONS EXERCISES:

Practice Set A

Write the decimal that goes with each written expression:

1) 4 and 5-thousandths ____________________

2) 345-thousandths ____________________

3) 11 and 2-hundredths ____________________

4) 1,234 and 56-hundredths ____________________

5) 5675-ten-thousandths ____________________

6) 1 and 1-tenth ____________________

7) 5-ten-thousandths ____________________

8) 4 and 1-millionth ____________________

9) 1,234,000 and 25-hundredths ____________________

10) 999 and 999-ten-thousandths ____________________

Write out how you would read the following numbers using the correct place values:

11) .009 ____________________

12) 2.12 ____________________

13) 159.5078 ____________________

14) .00034 ____________________

15) 798.099 ____________________

16) 49.9 ____________________

17) .499 ____________________

18) .0499 ____________________

19) .00499 ____________________

20) 25.6666 ____________________

21) In the number 1,286.0459,

a) what place does the 4 hold?

b) what place does the 0 hold?

c) what place does the 1 hold?

d) what place does the 9 hold?

22) In the number 798.0145603, name the places indicated by the following letters:

(arrows: b under the 8; a under the 4; c, d, e under the 6, 0, 3)

a) ____________

b) ____________

c) ____________

d) ____________

e) ____________

Give the EXPANDED FORM for each of the following decimals:

23) .125 ____________________

24) 23.55555 ____________________

25) .009 ____________________

26) .7 ____________________

27) 4000.024 ____________________

28) 4000.240 ____________________

29) 99.099 ____________________

30) .00062 ____________________

31) .8902 ____________________

32) 1,000,000,000.0001 ____________________

33) .000000005 ____________________

34) .0060 ____________________

35) 25.5 ____________________

Change the following decimals to FRACTION FORM (reduce answers wherever possible):

36) .375 ____________

37) .005 ____________

38) 11.75 ____________

39) 245.500 ____________

40) 245.5 ____________

41) 15.075 ____________

42) 9.625 ____________

43) .009 ____________

44) .008 ____________

45) .05 ____________

46) 1987.25 ____________

47) 10.5 ____________

48) 10.05 ____________

49) 8.453 ____________

50) .999 ____________

51) $24.\overline{3333}$ (a non-ending string of 3's) ____________

52) $66.\overline{6666}$ (a non-ending string of 6's) ____________

READING DECIMALS, DECIMALS TO EXPANDED FORM, AND DECIMALS TO FRACTIONS

Practice Set B

Write the decimal that goes with each written expression:

1) 3 and 2-hundredths ______

2) 125-thousandths ______

3) 15 and 35-thousandths ______

4) 1,487-hundred-thousandths ______

5) 7 and 7-thousandths ______

6) 226 and 50-hundredths ______

7) 226 and 5-hundredths ______

8) 545-millionths ______

9) 1 million and 1 millionth ______

10) 888 and 88-ten-thousandths ______

Write out how you would read the following numbers using the correct place value:

11) .008 ______

12) 4.56 ______

13) 4378.4193 ______

14) .0075 ______

15) 268.077 ______

16) 85.5 ______

17) 8.55 ______

18) .855 ______

19) 85.05 ______

20) 12.08333 ______

21) In the number 45,874.3201

a) what place value does the 1 hold?

b) what place value does the 5 hold?

c) what place value does the 0 hold?

d) what place value does the 3 hold?

22) In the number 865.130702
↑ ↑↑ ↑↑
c ba de

name the place values indicated by the following letters:

a) ________________

b) ________________

c) ________________

d) ________________

e) ________________

Give the EXPANDED FORM for each of the following decimals:

23) .375 ____

24) 4.99 ____

25) .0012 ____

26) .9 ____

27) 3450.028 ____

28) 3450.280 ____

29) 17.017 ____

30) .0054 ____

31) 2,002.0002 ____

32) .800007 ______________________

33) .0750 ______________________

34) 19.009 ______________________

35) .828 ______________________

Change the following decimals to FRACTION FORM (reduce answers wherever possible):

36) .625 ______________

37) .0025 ______________

38) 15.5 ______________

39) 15.005 ______________

40) 13.075 ______________

41) 3.375 ______________

42) 99.240 ______________

43) .007 ______________

44) .0016 ______________

45) 8.888 ______________

46) 40.6 ______________

47) 40.06 ______________

48) 2.805 ______________

49) $7.3\overline{33}3$ (a non-ending string of 3's) ______________

50) $25.6\overline{66}6$ (a non-ending string of 6's) ______________

EQUIVALENT DECIMALS AND ORDERING DECIMALS

The value of a decimal does not change if we attach zeros to the right of the last digit of the decimal. In effect we are adding 0 place values after the number.

Example 1: $.54 = .540 = .5400$, etc.

If we look at the expanded form of each of these decimals, they all involve $\frac{5}{10} + \frac{4}{100}$ and the rest are zeros. The decimals are all equal.

$$.54 = \frac{5}{10} + \frac{4}{100} \qquad .540 = \frac{5}{10} + \frac{4}{100} + \frac{0}{1000}$$

$$.5400 = \frac{5}{10} + \frac{4}{100} + \frac{0}{1000} + \frac{0}{10{,}000}$$

Example 2: $.85 = .850$ but $.85 \neq .805$

Let's look at the expanded forms.

$$.850 = \frac{8}{10} + \frac{5}{100} + \frac{0}{1000} = .85$$

$$.805 = \frac{8}{10} + \frac{0}{100} + \frac{5}{1000}$$

We can also see this if we read the two numbers.

.850 is read as 850-thousandths.

.805 is read as 805-thousandths.

They are different.

Example 3: $.026 \neq .260$

The expanded form of .026 is $\frac{0}{10} + \frac{2}{100} + \frac{6}{1000}$

The expanded form of .260 is $\frac{2}{10} + \frac{6}{100} + \frac{0}{1000}$

These expanded forms are *NOT* the same.

Also, .026 is read as 26-thousandths, while .260 is read as 260-thousandths.

They are different.

We get *equivalent decimal numbers* by putting any number of zeros to the *right* of the last digit of the decimal.

Therefore .26 = .260 = .2600 = etc.

but .26 ≠ .026 since the zero is to the *left*.

We can *order* decimals from smallest to largest, or vice versa, if we make each decimal have the same number of places to the right of the decimal point. Then all we need to do is compare the numbers, ignoring the point.

Example 4: List from smallest to largest: .456, .45, .4501, .4559

We make each decimal have the longest length of any of the numbers. In this example, that would be 4 decimal places after the point.

.456 = .4560
.45 = .4500
.4501 = .4501
.4559 = .4559

Reading the numbers without the point and ordering, they go 4500, 4501, 4559, and 4560. Therefore the smallest to largest are .45, .4501, .4559, .456.

This can be done for any set of decimal numbers smaller than 1. When there is a whole number part of the number, we must first determine which whole number part is largest or smallest.

Example 5: List from smallest to largest: 3.005, 3.0501, 4.10, 2.9, 2.99

The whole number parts are 2, 3, and 4.
The smallest whole number part is 2. Therefore, we compare 2.9 and 2.99.
We put both of these decimal numbers with two decimal places.
2.9 = 2.90 and 2.99 = 2.99.
The smallest is 2.90, then comes 2.99.

The next numbers involve the whole number part 3.
We put both the numbers to have 4 decimal places after the point.
3.005 = 3.0050 and 3.0501 = 3.0501. Then we compare 30,050 to 30,501.
The smaller is 3.005, then comes 3.0501.
That leaves the largest number 4.10.
In order the smallest to largest numbers are:
2.9, 2.99, 3.005, 3.0501, 4.10.

Another aspect of ordering decimal numbers is to try to place a particular number between two others.

Example 6: Between what two *whole numbers* does the decimal 2.59 lie?

If we tried to put this number on a number line, it would be between the whole numbers 2 and 3.

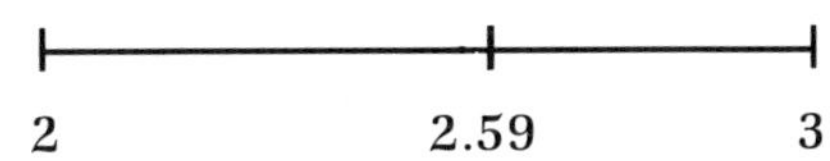

Example 7: Between what two numbers with *one decimal place* does the number 2.59 lie?

The number 2.59 lies between 2.5 and 2.6.

This was determined by looking at the *first decimal place*, 5, and seeing that our number is larger than that place, but less than the next place, 6.

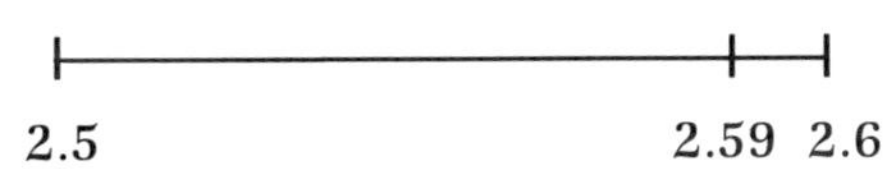

Since the number 2.59 has two decimal places, we only go as far as finding numbers around it up to 1 decimal place.

Example 8: Between what two numbers with *two decimal places* does the number 8.478 lie?

To determine this, we look at the *first two decimal places* after the decimal point of our number, 47. We use that number and the next consecutive number, 48.

Therefore, our number 8.478 lies between *8.47* and *8.48*.

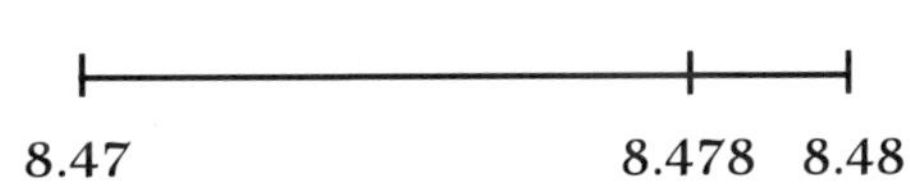

Example 9: Between what two numbers with *three decimal places* does the number 9.6543 lie?

Let's look at the first *three* decimal places after the decimal point of our number. Those places are 654. We use this one and the next consecutive number, 655. Our number, 9.6543 lies between 9.*654* and 9.*655*.

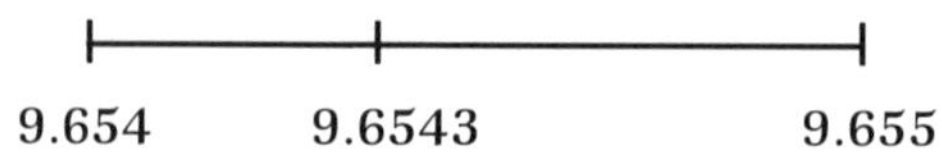

EQUIVALENT DECIMALS AND ORDERING DECIMALS EXERCISES:

Practice Set A

True or False:

1) .075 = .750

2) .230 = .2300

3) 4.99 = 4.999

4) .48 = .048

5) 6.3 = 6.300

6) 9 = 9.00

7) .60 = .6

8) 2.045 = 2.405

9) 12.02 = 12.020

Order the following decimals from smallest to largest:

10) 7.9901, 7.989, 7.9

11) .80, .88, .893, .8801, .9

12) 1.001, 1.0001, 1.0011

13) 1.90, 10.90, 10, 19.01, .190, 19.1

14) 3.03, 3.003, 3.0003, 3.031, 3.0035

15) 5.7, 5.77, 5.777, 5.79, 5.780, 5.7769, 5.76999, 5.799

Tell between what two WHOLE numbers the following decimals lie. Approximate its position on the number line between your two answers:

16) 18.765

17) 9.9604

18) .4507

19) .9999

20) 100.0005

21) 35.9999999999

Tell between what two numbers with ONE decimal place the following numbers lie. Place the numbers approximately on the number line between your two answers:

22) 29.678

23) 100.001

24) .4001

25) 99.9999

26) 69.99

27) .4999

28) .387

29) .025

30) .00765

31) .500002

Tell between what two numbers with TWO decimal places the following numbers lie. Place the numbers approximately on the number line between your two answers:

32) 29.678

33) 100.001

34) .4001

35) 99.9999

36) 69.995

37) .4999

38) .387

39) .025

40) .00765

41) .500002

42) .2389

43) .145

44) 100.055

45) 8.362

46) .997

Tell between what two numbers with THREE decimal places the following numbers lie. Place them approximately on the number line between your two answers:

47) 8.76543

48) .010475

49) 4.499999

50) 25.00049959

EQUIVALENT DECIMALS AND ORDERING DECIMALS EXERCISES:

Practice Set B

True or False:

1) .0025 = .0250
2) .48 = .488
3) 5.51 = 5.510
4) 6.302 = 6.320
5) 1.7 = 1.700
6) .709 = .79
7) .333 = .3
8) 19.99 = 19.990
9) 3.004999 = 3.005

Order the following decimals from smallest to largest:

10) 10.566, 10.57, 10.5669, 10.56689
11) .30, .33, .3301, .329
12) 9786.79, 9786.789, 9786.79001
13) 8.9, 8.99, 8.989, 8.91, 8.9909, 8.9099
14) 5.001, 5.0009, 5.01, 5.1, 5.00989

Tell between what two WHOLE numbers the following decimals lie. For each decimal, place it on the number line between your two answers:

15) 75.033
16) 19.989
17) 200.01
18) 91.999
19) 229.0002

Tell between what numbers with ONE decimal place the following numbers lie. Place the answers approximately on the number line between your two answers:

20) 35.275
21) 89.019
22) .650001
23) .0034
24) .601
25) .39999

Tell between what two numbers with TWO decimal places the following numbers lie. Place the numbers approximately on the number line between your two answers:

26) 35.275

27) 89.019

28) .650001

29) .0034

30) .601

31) .39999

32) 128.325

33) 9.9901

34) 4.09059

Tell between what two numbers with THREE decimal places the following numbers lie. Place them approximately on the number line between your two answers:

35) .00989

36) 9.9901

37) 4.09059

38) .01275

39) .4459

40) 256.083315

ADDITION AND SUBTRACTION OF DECIMALS

Since decimals are really fractions with denominators that are powers of 10, we can add or subtract them only if they have the same denominators. With decimal forms, this means that they must have the *same number* of places after the decimal point.

Example 1: Add .6 + .25

Since $.6 = \frac{6}{10}$ and $.25 = \frac{25}{100}$, we can only add them if both are in hundredths. So we must put the .6 to its equivalent decimal form with two places after the decimal point. (.6 = .60)

It helps to add the decimals vertically:

$$\begin{array}{r} .60 \\ +\,.25 \\ \hline .85 \end{array} \quad \text{is the answer}$$

> RULE: To ADD numbers, it helps to set the problem up vertically. Make sure the decimal points fall in the same vertical column. Make each number have the same number of decimal places after the decimal point by adding zeros to the right as needed. Add the digits having the same positonal value. Bring down the decimal point in the answer.

Example 2: Add 45 + 4.5 + .45 + .045

We set this up vertically. Each number needs three decimal places (because .045 has 3 decimal places)

$$\begin{array}{r} 45.000 \\ 4.500 \\ .450 \\ +\quad .045 \\ \hline 49.995 \end{array}$$

Example 3: Add 100.1 + 11.01 + 100.001 + 110.

Again, we set this problem up vertically. Since the number 100.001 has 3 decimal places, we make each number have 3 decimal places.

$$\begin{array}{r} 100.100 \\ 11.010 \\ 100.001 \\ +\,110.000 \\ \hline 321.111 \end{array} \qquad \text{The answer is } 321.111$$

> RULE: To SUBTRACT numbers, the decimal points must be in the same vertical column. Each number should have the same number of places after the decimal point. Subtract digits with the same positional value. Bring down the decimal point in the answer.

Example 4: Subtract 9.874 − 3.05

It is easier to set this problem up vertically.

$$\begin{array}{r} 9.874 \\ -\,3.050 \\ \hline 6.824 \end{array} \qquad \text{is the answer.}$$

Example 5: 10 − 4.973

Once again we set this up vertically. Each number will have 3 places after the decimal point.

$$\begin{array}{r} 10.000 \\ -\quad 4.973 \\ \hline \end{array} \qquad \text{We need to borrow from the 10.} \qquad \begin{array}{r} \overset{9\ \ 9\ 9\ 10}{\cancel{1}\cancel{0}.\cancel{0}\cancel{0}\cancel{0}} \\ -\quad 4.973 \\ \hline 5.027 \end{array}$$

The answer is 5.027.

Example 6: Compute .075 + 2 − 1.9786

First we add .075 + 2.

$$\begin{array}{r} .075 \\ +\,2.000 \\ \hline 2.075 \end{array}$$

Then we subtract.

$$2.075 - 1.9786 = \begin{array}{r} 2.0750 \\ -\,1.9786 \\ \hline \end{array} = \begin{array}{r} \overset{1\ \ 9\ 16\ 14\ 10}{\cancel{2}.\cancel{0}\cancel{7}\cancel{5}\cancel{0}} \\ -\,1.9786 \\ \hline .0964 \end{array}$$

The answer is .0964.

ADDITION AND SUBTRACTION OF DECIMALS EXERCISES:

Practice Set A

Add and/or subtract as the problem dictates. It helps to set the problems up vertically in most instances. Try not to use the calculator. Do parentheses first.

1) 3.25 + .004 + .1250 =

2) 119.8
.03
39
+ 4.0004

3) 88.8 + 88.08 + 88.008 =

4) 112 − 93.45 =

5) $2345.98
− $1299.49

6) .0000000789 + .00003 + .00127 =

7) 10 − (1.25 + 3.375)

8) 4 + .125 + 40 − (36 − .55)

9) .67 + .669 + .6701 + .067 =

10) .1 + .10 + .01 + .001 + .100 + .0001 + .1000 =

11) .00205
34.66
6.559
+ 85.68088

12) $ 46.98
$ 23.009
$ 50.05
+ $4467.95

13) 15 − 3.981

14) .8903
− .0975

15) 8 − 1.75

16) $45.60
− 19.75

17) 100 − 24.999 =

18) 15 − (.025 + 3.7051) =

19) $512.23 − ($156.75 − $19.98)

20) 1 + .2 + .03 + .004

ADDITION AND SUBTRACTION OF DECIMALS EXERCISES:

Practice Set B

Add or subtract as the problem dictates. Try not to use the calculator.

1) 4.83 + .0025 + .17063

2)
```
   315.3
      .005
   400
+    2.00006
```

3) 99.9 + 99.09 + 99.009

4)
```
  $45,163.87
+ $ 1,243.79
```

5) 2001 − 1099.098

6) .00000258 + .0006 + .00784

7) 25 − (2.85 + 16.176)

8) 15 + .375 + 30 − (41 − .98)

9) .58 + .558 + .5801 + .058

10) .3 + .30 + .03 + .003 + .300 + .0003 + 3

11)
```
     .00716
   38.2
   90
+    .650008
```

12)
```
  $ 25.79
  $ 46.05
  $   4.98
+ $258.005
```

13) 18 − 9.099

14)
```
   .00763
 − .00694
```

15) 256 − 4.56

16) 2000 − 1798.0889

17) 23 − (14.03 + .1595)

18) $2456.78 − ($1005.41 + $700)

19) .3 + .5 + .7 + .09

20) 10,000 − 999.887

APPLICATIONS OF ADDITION AND SUBTRACTION OF DECIMALS EXERCISES:

Practice Set A

1) Andy had a balance of $105.75 in his checking account. During the month he wrote checks as follows, $23.6, $5.75, $84.85, and $2.60. He received a note from his friendly bank saying, you are overdrawn by this amount, ______________________.

2) The adjoining figure shows the cross section of a culvert. Find its thickness.

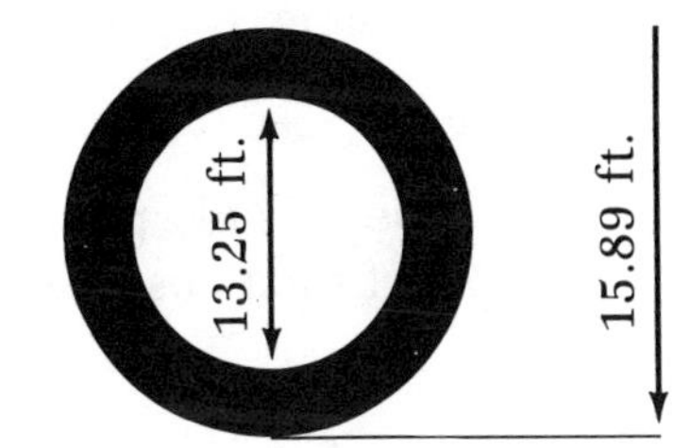

3) A man has 240.5 ft. of fencing. How close would he come to being able to enclose this field?

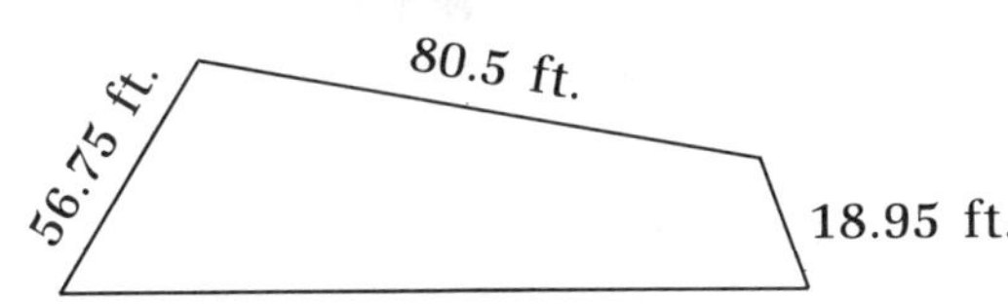

4) A charitable organization received gifts of $150.75, $28.98, $15.00 and 75 cents. If they previously had a balance of $2763.19 in their checking account, what would be their new balance if they deposited all the gifts?

5) Your monthly PG&E bill arrives and you see that this month you used 24.3 kilowatt hours per day as compared to last year during the same month when you used 19.9 kilowatt hours per day. What is the difference between the two usages?

6) The activity in your checking account for the past week looks like this: Monday, deposited $150.00; Tuesday, wrote check for $23.78; Wednesday, wrote checks for $14.16, $11.85, and $50.00; Thursday, withdrew $100 in cash; Friday, deposited salary check for $724.36.

   If the balance is in your checking account at the beginning of the week was $423.19, what is it at the end of the week?

7) How much larger is 17.83 than 17.38?

8) The fertility rate in the United States is 61.1 per 1000 females. That means that on the average 61.1 women out of every 1000 females give birth each year. The fertility rate in the People's Republic of China is 112.5 per 1000 females. How many more babies are born per 1000 females in China as compared to the United States?

9) As of 1984, the population of dogs in the United States was estimated at 50.2 million. That same year, the population of cats was estimated at 46.3 million. In 1982, the population of cats was only 23 million.

a) How many more dogs than cats were there in the United States in 1984.

b) How large was the increase in cats from 1981 till 1984?

*10) A bottle of soda pop and a cup of ice to drink it from together cost $1.10. If the bottle of soda costs $1.00 more than the cup of ice, how much does each cost?

*11) A young boy enters a store and tells the owner that if he (the owner) gives him as much money as he (the young boy) has in his pocket, he will spend $10 in the store. The owner agrees and the boy spends the money.

The young boy enters the second store and tells that owner the same thing. The owner complies and the boy spends the $10.

The boy enters a third store. He tells the owner the same thing. That owner also agrees, the boy spends the $10 but has NO money left to try out in another store.

How much money did the young boy have to start with?

APPLICATIONS OF ADDITION AND SUBTRACTION OF DECIMALS

Practice Set B

1) A man has 300.5 feet of fencing. Would he have enough to enclose this field?

   If not, how much would he need?

   If he has enough, how much would he have left over?

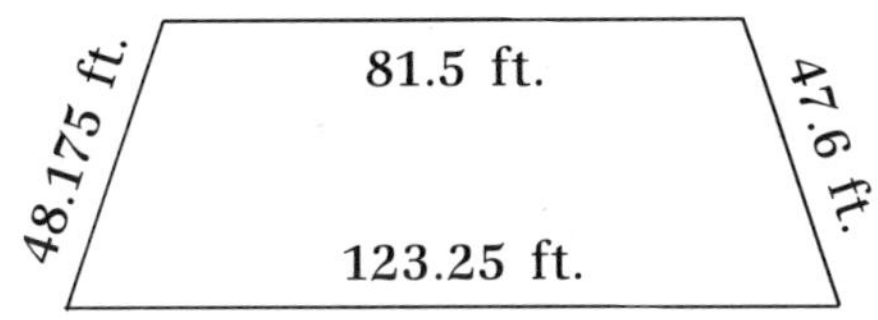

2) Your checking account statement for July arrives in the mail. It shows that your previous month's balance was $1836.78. During July you made 5 deposits totalling $520.48; you wrote 30 checks totalling $2147.62; there was a monthly service charge of $1.08; and there was a charge for new checks of $13.45. What current balance should your July statement show?

3) At the beginning of a long car trip touring the United States your car's odometer reads 39,674.5 miles. By the time you return home your car's odometer reads 43,271.1 miles. How many miles did you drive in your car?

4) Raunchy Ray, a sly antique dealer, sold a certain brooch for $925.75, bought the same brooch back for $879.97, and then resold the brooch for $1002.24. What did he make on the transaction?

5) A contractor is measuring a bedroom wall in preparation for the installation of a new window. The rough measurement from the floor of the room to the top of the window opening is 78.25 inches. The measurement from the floor to the bottom of the window opening is 32.875 inches. How long (vertically) is the window opening?

6) You own 1000 shares in True Blue Mutual Fund company. You want to be invested only in mutual funds that don't vary more than $3.15 week by week (this is what you consider a STABLE fund). In watching the activity of your fund in the Wall Street Journal you see that one week the fund's closing price as of Friday was $52.47. In three consecutive Fridays the closing price was $54.61, $51.99 and $47.68. Is this mutual fund stable enough for you?

7) How much larger than 9.879 is 9.987?

8) A color television set is advertised at $499.99. The final price including tax is $531.44. How much is the sales tax?

9) The highest fertility rate in the world is in the African country south of Algeria called Mali. On the average, there are 262.4 women who give birth per 1000 females in the population.

The lowest fertility rate is in Monaco, the independent monarchy on the French Riviera. There are only 25.5 women per 1000 in the population that give birth.

How many more women give birth per 1000 in Mali than in Monaco?

*10) How can you have seven coins totalling $1.07 but with these coins you couldn't give change for a half-dollar or a quarter or a dime or a nickel?

MULTIPLICATION AND DIVISION BY POWERS OF 10

Our number system is constructed around powers of 10. That is why it is called a base-10 system. If we review positional values of our number system, we see that they are increased by a factor of 10 as we move to the left in our place value columns.

$$\text{etc.}\quad \underbrace{10 \times 10 \times 10 \times 10}_{= 10{,}000} \quad \underbrace{10 \times 10 \times 10}_{= 1000} \quad \underbrace{10 \times 10}_{= 100} \quad \underbrace{10}_{= 10} \quad 1 = 1 \quad \longleftarrow$$

The numbers are multiplied by 10 for every step to the left. ←

Multiplying by 10 puts another zero in the number.

$$\text{etc.} \xleftarrow{} 10000 \xleftarrow{\times 10} 1000 \xleftarrow{\times 10} 100 \xleftarrow{\times 10} 10 \xleftarrow{\times 10} 1$$

As we move to the right in our place value columns, the positional values are divided by 10. →

$$100 \xrightarrow{\div 10} 10 \xrightarrow{\div 10} 1 \xrightarrow{\div 10} \frac{1}{10} \xrightarrow{\div 10} \frac{1}{100} \xrightarrow{\div 10} \frac{1}{1000} \xrightarrow{\div 10} \frac{1}{10{,}000} \quad \text{etc.}$$

(Recall that 1 divided by 10 $= 1 \div 10 = \frac{1}{10}$)

Dividing by 10 puts another zero in the denominator.

$$\frac{1}{1}, \frac{1}{10}, \frac{1}{100}, \frac{1}{1000}, \text{ etc.}$$

Let's look at decimal numbers to see how the values change as we multiply and divide by powers of 10.

Example 1: $.75 \times 10$

Changing this problem to fraction form will show us a shortcut.

$$.75 \times 10 = \frac{75}{100} \times 10 = \frac{75}{\cancel{100}_{10}} \times \cancel{10}^{1} = \frac{75}{10} = 7\frac{5}{10} = 7.5$$

The answer is 7.5.

We started with the number .75, multiplied by 10, and got the answer 7.5.

The effect of multiplying by 10 was to move the decimal point ONE PLACE TO THE RIGHT.

$$.75 \times 10 = \underset{\times 10}{.75} = 7.5$$

Example 2: $.873 \times 100$

Let's look at this problem in fractions.

$$.873 \times 100 = \frac{873}{1000} \times 100 = \frac{873}{\underset{10}{\cancel{1000}}} \times \overset{1}{\cancel{100}} = \frac{873}{10} = 87\frac{3}{10} = 87.3$$

We started with the number .873, multiplied by 100, and got the answer 87.3.

The effect of multiplying by 100 was to move the decimal point TWO PLACES TO THE RIGHT.

$$100 = .\underbrace{8\ 7}_{\times 100}\ 3 = 87.3$$

Now we can see the following shortcut:

MULTIPLYING A NUMBER BY A POWER OF 10, WE MOVE THE DECIMAL POINT AS MANY PLACES TO THE *RIGHT* AS THE NUMBER OF ZEROS IN THE POWER OF 10.

Example 3: 3.4507×1000

There are *three* zeros in the power of 10.

We move the decimal point *three* places to the *right*. $3.\underbrace{4\ 5\ 0}_{\times 1000}\ 7 = 3450.7$

The answer is 3450.7.

Example 4: 25×100

It is understood that the decimal point is after the 25.

We put zeros as place holders after the point.

$25. = 25.00$

Now we move the decimal point *two* places to the *right*. $25.\underbrace{0\ 0}_{\times 100} = 2500.$

The answer is 2500.

(We have seen this type of problem before. On our Shortcut to Mental Addition Worksheet, we saw to attach zeros at the end of the number.)

Now we will look at dividing by powers of 10.

Example 5: $25 \div 10$

Translating this directly into fraction form, it is

$$25 \div 10 = \frac{25}{10} = 2\frac{5}{10} = 2.5$$

We started with 25, divided by 10, and got the answer 2.5.

The effect of dividing by 10 was to move the decimal point ONE PLACE TO THE LEFT.

$$25. \div 10 = 2\,5\,. = 2.5$$

$\div 10$

Example 6: $78 \div 100$

Translating this directly into fraction form, we get

$$78 \div 100 = \frac{78}{100} = .78$$

We started with the number 78, divided by 100, and got the answer .78.

$$78. \div 100 = 7\,8\,. = .78$$

$\div 100$

The effect of dividing by 100 was to move the decimal point TWO PLACES TO THE LEFT.

Now we have a shortcut.

DIVIDING A NUMBER BY A POWER OF 10, WE MOVE THE DECIMAL POINT AS MANY PLACES TO THE *LEFT* AS THE NUMBER OF ZEROS IN THE POWER OF 10.

Example 7: $84.6 \div 100$

There are *two* zeros in the power of 10.

We will move the decimal point *two* places to the left.

$$84.6 \div 100 = 8\,4\,.6 = .846$$

$\div 100$

The answer is .846.

Example 8: $.64 \div 100$

There are *two* zeros in the power of 10.

We will move the decimal point *two* places to the left.

First we need to put zeros as place holders in front of the decimal point.

$.64 = 00.64$

Now we move the decimal point *two* places to the left.

$00.64 \div 100 = 00.64 = .0064$

The answer is .0064.

MULTIPLICATION AND DIVISION BY POWERS OF 10 EXERCISES:

Practice Set A

Multiply by these powers of 10:

1) 34 × 100
2) .34 × 100
3) 148 × 1000
4) 1.48 × 1000
5) .148 × 1000
6) 8.00 × 10
7) .800 × 10
8) .0008 × 10
9) 12.5 × 10
10) 30 × 100
11) .1250 × 10,000
12) 1.25 × 1000
13) 11 × 1000
14) .11 × 1000
15) 1.1 × 100
16) 175.8 × 100
17) 2.375 × 10
18) .0625 × 1000
19) 1 × 100,000
20) .1 × 10,000

Divide by these powers of 10:

21) 11 ÷ 10
22) 1.1 ÷ 10
23) 124 ÷ 10
24) .124 ÷ 100
25) 769 ÷ 100
26) 769 ÷ 1000
27) 76.9 ÷ 10
28) 76.9 ÷ 1000
29) .769 ÷ 10
30) .0769 ÷ 100
31) 4 ÷ 10
32) .4 ÷ 100
33) 100 ÷ 10
34) .1 ÷ 100
35) 10 ÷ 100
36) 95 ÷ 1000
37) 1625 ÷ 10,000
38) .073 ÷ 100
39) 152 ÷ 10
40) 38.9 ÷ 10

In each of the following exercises, fill in the blank with the correct operation (× or ÷) and the correct power of 10. For example: 4.57 __________ = 45.7
The correct solution is *× 10*. 4.57 *× 10* = 45.7

41) 186 __________ = 18,600
42) 96.3 __________ = .0963
43) 13 __________ = .0013
44) 75 __________ = .75
45) .025 __________ = 2.5
46) 8.4 __________ = 8400

47) 2.80 __________ = .028

48) 24.5 __________ = 2.45

49) 100 __________ = .10

50) .10 __________ = 100

51) Fill in the following table. The column headings should tell whether to multiply or divide by a power of 10. Figure out which by the entries given:

#	?	?	?	?
24	2400			
18		.18		
7.5				75
1			.001	
.4	40			
30			.03	
6.8				68
9		.09		

? = multiply or divide by the appropriate power of 10.

MULTIPLICATION AND DIVISION BY POWERS OF 10 EXERCISES

Practice Set B

Multiply by these powers of 10:

1) 48 × 100
2) .48 × 100
3) 263 × 1000
4) 2.63 × 1000
5) .263 × 1000
6) 7.10 × 10
7) .710 × 10
8) .0007 × 100
9) 15.75 × 10
10) 45 × 100
11) .3750 × 10,000
12) 6.275 × 1000
13) 12 × 1000
14) .12 × 1000
15) 2.2 × 100
16) 298.99 × 1000
17) 3.408 × 10
18) .0025 × 1000
19) .01 × 10,000
20) .1 × 100,000

Divide by these powers of 10:

21) 15 ÷ 100
22) 5.6 ÷ 10
23) 248 ÷ 10
24) .248 ÷ 100
25) 4078 ÷ 100
26) 4078 ÷ 1000
27) 40.78 ÷ 10
28) 40.78 ÷ 1000
29) 4.078 ÷ 10
30) .04078 ÷ 100
31) 6 ÷ 10
32) .06 ÷ 100
33) 100 ÷ 1000
34) .01 ÷ 10
35) 72 ÷ 1000
36) 2431 ÷ 100
37) 2.436 ÷ 1000
38) .095 ÷ 10
39) 342 ÷ 100
40) 57.9 ÷ 10

In each of the following exercises, fill in the blank with the correct operation (× or ÷) and the correct power of 10.

41) 275 __________ = 2750
42) 654.3 __________ = 6.543
43) 27 __________ = .027
44) 83 __________ = 8.3
45) .125 __________ = 1250
46) 9.2 __________ = 920

47) 3.00 __________ = 3

48) 19.98 __________ = 199.8

49) 2250 __________ = .225

50) .0300 __________ = 30

51) Fill in the following table. Complete the column headings to tell whether to multiply or divide by a power of 10. Figure out which by the entries given:

#	?	?	?	?
46				46000
15	.015			
8.7		87		
5			.05	
.3				300
60	.06			
2.1		21		
.89			.0089	

SCIENTIFIC NOTATION

Each morning we lift ourselves out of bed, and this requires energy. At the time, the amount of energy seems very large, but it is small when compared to a minor earthquake of 1 on the Richter scale. To relate these amounts of energy we use the basic unit of the "erg". An erg is not very much. For example, an "erg" is the amount of energy needed to lift one gram (the weight of a dime) up one centimeter (about $\frac{2}{5}$ of an inch). So it might require 1,000,000 ergs to get out of bed in the morning, but a lightning bolt strikes with 10,000,000,000,000,000 ergs.

To help us deal with such large numbers, we have developed a shorthand notation for writing these numbers. It is called SCIENTIFIC NOTATION.

When we work with large numbers it can be very tedious counting all the zeros. Can you tell at a glance the difference between 300000000 and 3000000000?

Scientific Notation is an exponential form which makes it easy for us to keep track of all the zeros.

> SCIENTIFIC NOTATION expresses any number as the *product* of a number "n" between 1 and 10, and a power of 10.
>
> $$n \times 10^k$$

300000000 $= 3 \times 10^8$ in scientific notation (since there are 8 zeros).

3000000000 $= 3 \times 10^9$ in scientific notation (since there are 9 zeros).

> RULE: To express any number in scientific notation, count the *number of places* you must move the decimal point in the original number to be behind the first nonzero digit.
>
> Scientific Notation is the new resulting number (between 1 and 10), TIMES 10 to the power of the counted places.

Example 1: Express 4,250,000,000,000,000 in scientific notation.

$$4,250,000,000,000,000 = 4.25 \times 10^{15}$$

15 12 9 6 3

The exponent is 15 since we counted 15 places to move the decimal point behind 4.

Example 2: Express 876,100,000 in scientific notation.

8 . 7 6 1 0 0 0 0 0 = 8.761×10^8 (There were 8 places moved.)

8 7 6 5 4 3 2 1 places

Example 3: Express 78.9 in scientific notation.

7 8 . 9 = 7.89×10^1 (The exponent is 1 because we moved 1 place.)

1 place

Example 4: Express 1586.70 in scientific notation.

1 5 8 6 . 7 0 = 1.5867×10^3 (The exponent is 3 because we moved the decimal point 3 places.)

3 places

We can also express a number, given in scientific notation, in STANDARD FORM. This involves multiplying by a power of 10.

Example 5: Change 4.83×10^3 to standard form.

$4.83 \times 10^3 = 4.83 \times 1000 =$ 4 . 8 3 0 = 4830.

(Recall from a previous section that multiplying by 1000 involves moving the decimal point 3 places to the right.)

Example 6: Change $.0067 \times 10^6$ to standard form.

$.0067 \times 10^6 = .0067 \times 1000000 =$. 0 0 6 7 0 0 = 6700.

Multiplying by 10^6 moves the decimal point 6 places to the right.

NOTE: All powers of 10 in these numbers larger than 1 are *positive*. Algebra classes will give meaning to negative powers of 10.

SCIENTIFIC NOTATION EXERCISES:

Practice Set A

Write each of the following numbers in scientific notation:

1) 170 2) 4500 3) 1020 4) 65,000 5) 29,100,000

6) 8,015,000,000,000 7) 167 8) 5 9) 28

10) 34.8 11) 805,000 12) 798,000,000,000,000 13) 498.16

14) 152.3 15) 19.5 16) 7.8 17) 53,800 18) 674

19) 111,010,000 20) 9,990,000,000,000,000,000,000

Write each of the following numbers in standard form:

21) 3.7×10^2 22) 4.65×10^3 23) 1.01×10^6

24) 7.98×10^1 25) 9.56×10^4 26) 5.0046×10^{20}

27) $.08 \times 10^5$ 28) $.004 \times 10^6$ 29) 7000×10^3

30) 7×10^0 31) 19×10^1 32) 4.16×10^{14}

33) 1×10^{10} 34) $.2 \times 10^4$ 35) $.06 \times 10^4$

36) 5.8×10^{11} 37) $.00153 \times 10^{15}$ 38) 9.9×10^{16}

39) 4.1×10^3 40) $.0075 \times 10^6$

Write each number that appears in the following statements in scientific notation:

41) Did you know that the estimated age of the universe is 320,000,000,000,000,000 seconds?

42) The earth completes one orbit around the sun in approximately 31,500,000 seconds.

43) The sun is a star whose diameter measures approximately 869,000 miles.

44) The greatest thickness of ice ever measured was in Antarctica. The thickness was 14,600 feet.

45) Sunspots are 100,000 times as bright as the full moon though they appear dark against the sun's glare.

46) If Antarctica's 7,000,000 cubic miles of ice were to melt, it would raise the level of the ocean by 180 feet.

47) Light travels at the rate of 300,000,000 meters each second. A "light year" is the distance that light would travel in one year (assuming 365 days).

a) How many meters would light travel in 1 minute?

b) in 1 hour?

c) in 1 day?

d) in 1 year?

Therefore, a light year is __________ meters.

e) The star closest to the sun is called "Proxima Centauri" and is only 4.3 light years away from the sun. How many meters is that?

SCIENTIFIC NOTATION EXERCISES:

Practice Set B

Write each of the following numbers in scientific notation:

1) 340 2) 7800 3) 2010 4) 47,000 5) 38,500,000

6) 9,456,000,000,000 7) 238 8) 6 9) 37

10) 13.7 11) 916,000 12) 234,000,000,000,000,000,000

13) 586.87 14) 456.8 15) 21.8 16) 8.3 17) 37,200

18) 723 19) 1,011,000,000 20) 4,096,000,000,000,000

Write each of the following numbers in standard form:

21) 4.6×10^2 22) 8.39×10^3 23) 2.11×10^6

24) 6.77×10^1 25) 9.19×10^4 26) 2.1103×10^{16}

27) $.0063 \times 10^5$ 28) $.104 \times 10^6$ 29) 4500×10^3

30) 8×10^0 31) 23×10^1 32) 8.81×10^{13}

33) 4×10^0 34) $.6 \times 10^3$ 35) $.09 \times 10^4$

36) 7.7×10^{11} 37) $.00463 \times 10^{12}$ 38) 99×10^{10}

39) 1×10^{11} 40) $.125 \times 10^5$

Write each number that appears in the following statements in scientific notation:

41) A large oak tree can have 250,000 leaves, but a cypress can have as many as 50 million leaf scales.

42) Seeds to grass pollen can number 6 billion per ounce.

43) The largest wooded area exists in the northern part of the USSR. The amount of wooded area is 2,700 million acres.

44) Pluto is the planet most distant from the Sun. Its maximum distance from the sun is 4,600,000,000 miles.

45) As of 1985, the population of the world was 4,865,000,000. Of this total, the number of females was 2,443,000,000. How many males?

46) The Soviet Union is the country with the largest total area. It occupies 8,649,000 square miles. The United States is the fourth largest country by area, occupying 3,683,000 square miles.

47) The busiest subway system is in the Greater Moscow area. It services as many as $6\frac{1}{2}$ million passengers per day.

MULTIPLICATION OF DECIMALS

By now we see that the quantity of decimal places in a decimal number corresponds directly with the name of the number. For example, the decimal .645 has *three* decimal places, and its name is *thousandths*. The *three* decimal places correspond directly with the *three* zeros in the denominator of thousandths ($\frac{1}{1000}$).

The decimal number 18.09 has *two* decimal places, and its name is *hundredths*. The *two* decimal places is the same as the *two* zeros in hundredths ($\frac{1}{100}$).

Since decimal numbers are really fractions with denominators that are powers of 10, when we multiply them we are really multiplying fractions. We multiply numerators separately, and denominators separately. Since the denominators are just powers of 10, we are just dealing with attaching zeros in the denominator of the answer.

Example 1: .25 × .85

Let's do this problem first in fractions. It will show us the shortcut.

$$.25 \times .85 = \frac{25}{100} \times \frac{85}{100} = \frac{25 \times 85}{10{,}000} = \frac{2125}{10{,}000} = .2125$$

Note how our final fraction has *four* zeros in the denominator.
That means that our answer has *four* decimal places.
The numerator of the final fraction is just the product of 25 and 85.
(25 × 85 = 2125)

The shortcut with this problem is to multiply the numbers, disregarding the decimal places. In the final result, count the total number of places in the problem. The final result should have *that* number of decimal places. (4 places in this problem.)

$$.25 \times .85 = .2125 \qquad \text{(think } 2125 = 25 \times 85\text{)}$$

> RULE: When multiplying two decimal numbers, multiply the numbers as if there were no decimal point. After you have that number, count the *total* number of decimal places in the problem. The final answer should have that number of places. Count left from the result.

Example 2: 1.45 × 6

First we multiply 145 × 6 to get 870. There is a total of *two* decimal places in the problem. The final answer is 8.70 with two decimal places. (8 7 0 .)
(Note: 8.70 is the same as 8.7. They are equivalent forms.)

Example 3: 1.9 × .19

First we multiply 19 × 19 = 361.

Counting decimal places in the problem, there is a total of *three* places. Our answer should have *three* places. The answer will be gotten by moving *three* places back from the 1 (. 3 6 1) The final answer is .361.

Example 4: .75 × .004

First we multiply 75 × 4 which is 300.

Our answer should have *five* decimal places. We count *five* places left from the end of the number 300. Our final answer is .00300. (. 0 0 3 0 0)

(NOTE: We can ignore zeros at the end of a decimal number. .00300 = .003).

Example 5: .03 × .005 × .20

Disregarding the decimal points, we are multiplying 3, 5 and 20. We multiply out 3 × 5 × 20 to get 300. Our answer should have *seven* decimal places. Counting left from the end of 300, our final answer is .0000300. (. 0 0 0 0 3 0 0)

(NOTE: .0000300 = .000030 = .00003)

(RECALL: Zeros at the end can be dropped, but not at the beginning. .0000300 = .000030 = .00003)

MULTIPLICATION OF DECIMALS EXERCISES:

Practice Set A

In each of the following exercises, put the decimal point in the correct place. The numeric parts of the answers are correct:

1) 1.25 × .5 = 6 2 5
2) .18 × 3.4 = 6 1 2
3) 12.1 × .009 = 1 0 8 9
4) .0064 × 6.4 = 4 0 9 6
5) 891 × .025 = 2 2 2 7 5
6) .2 × .30 × .006 = 3 6 0
7) 12.98 × .46 = 5 9 7 0 8
8) .4 × .08 × .16 = 5 1 2
9) $(1.2)^3$ = 1 7 2 8
10) 3200 × .005 = 1 6 0 0 0

Multiply out the following without using the calculator:

11) .25 × 3
12) 20 × .5
13) .8 × .03 × .40
14) 2.1 × .021
15) 8.4 × .025
16) .005 × .0006 × 7.000
17) 2.200 × 2.2
18) $(.2)^4$
19) $(1.5)^2$
20) $(.03)^4$
21) $(.5)^3$
22) 20.5 × 1.5
23) 1.1 × .11
24) .138 × 5.5
25) 1.5 × .015
26) .024 × 2.4
27) .33 × .004
28) 8.5 × .6
29) 7.55 × .023
30) 23.2 × .09

MULTIPLICATION OF DECIMALS EXERCISES:

Practice Set B

In each of the following exercises, put the decimal point in the correct place. The numeric parts of the answers are correct:

1) 1.6 × 2 = 3 2
2) .045 × 2 = 9 0
3) 2.03 × 3.0 = 6 0 9 0
4) 1.45 × 6 = 8 7 0
5) 14.78 × .235 = 3 4 7 3 3 0
6) 5.6 × 12.9 = 7 2 2 4
7) .096 × 23 = 2 2 0 8
8) 4.8 × .25 = 1 2 0 0
9) .0036 × 3.6 = 1 2 9 6
10) $(3.2)^2$ = 1 0 2 4

Multiply out the following without using the calculator:

11) .006 × 4.8
12) 75 × .04
13) .12 × 3 × .04
14) 1.3 × .013
15) 6.4 × 12
16) .003 × .04 × .8
17) 5.5 × .550
18) $(1.9)^2$
19) $(.4)^4$
20) $(2.3)^3$
21) $(.025)^2$
22) 17.2 × 3.4
23) 1.7 × .017
24) 17.2 × 2.4
25) 1.8 × .0018
26) 36 × .004
27) 1.001 × 2.5
28) 1.25 × .33
29) 7000 × .15
30) 8.8 × 2.02

DIVISION OF DECIMALS

We can divide decimals as easily as we multiply them if we ALWAYS MAKE THE DIVISOR A WHOLE NUMBER. When the divisor is not a decimal, we can divide as we do normally. The result will have the same number of decimal places as the original dividend.

Let's look at an example. We will set the problem up first with fractions. This will justify what steps we are taking.

Example 1: Divide .32 ÷ 8

We set this problem up with fractions.

$$.32 \div 8 = \frac{32}{100} \div \frac{8}{1} = \frac{32}{100} \times \frac{1}{8} = \frac{32}{8} \times \frac{1}{100} = 4 \times \frac{1}{100} = \frac{4}{100} = .04$$

The answer is .04. It has the same number of decimal places as the original dividend (.32).

Because we were dealing with a whole number divisor, 8, we were able to keep the hundredths. That is why the answer has 2 decimal places. ($\frac{1}{100}$ has 2 zeros.)

We do not have to go through this fraction problem to arrive at the correct answer. We simply divide 8 into .32 as if we were working with numbers without decimal points.

$$8\overline{).32}\quad = .04$$

The decimal point stays in the same place throughout the problem.

> RULE 1: When dividing a decimal number by a *whole number*, we divide the numbers as if there were *no* decimal point. The final answer will have the same number of decimal places as appears in the problem.

Example 2: Divide .45 ÷ 15

We work this problem as if there were no decimal point.

$$15\overline{)45}\quad = 3$$ The result is 3.

Our answer will be .03. Our answer will have TWO decimal places since the dividend .45 has TWO decimal places.

Another approach is to divide directly.

$$15\overline{).45}\quad \text{quotient } .03$$

The decimal point stays in the same place throughout the calculation.

Example 3: Divide $27\overline{).405}$

Working this problem directly, we divide:

$$27\overline{).405}\quad \text{quotient } .015$$

The answer is .015.

When division problems have divisors that are NOT whole numbers, we must first change those divisors to BE whole numbers. Then we can proceed to divide as we just did. Since it is relatively easy to divide a decimal by a whole number, we will change all division problems to this form. This can be done much the same way we worked with equivalent fractions.

We will think of every division problem as if it were in fraction form. The numerator is the *dividend*, and the denominator is the *divisor*.

We will multiply the decimal denominator (since it is the DIVISOR) by a power of 10 sufficient to make it a whole number. To keep the equivalence in the problem, we will multiply the numerator by the same power of 10.

Example 4: $.49 \div .07$

We can write this problem in fraction form as $\frac{.49}{.07}$.

We need to multiply the denominator by 100 to make it a whole number. We will do the same to the numerator. (This doesn't change the value of the fraction.)

$$\frac{.49}{.07} \underset{\times 100}{\overset{\times 100}{=}} \frac{49}{7}$$

When we multiply by 100 we are moving the decimal point in the denominator TWO places to the RIGHT. We must also move the decimal point in the numerator TWO places to the RIGHT.

$$.07\overline{).49} = .07\overline{).49} = 7\overline{)49} \quad \text{(quotient } 7\text{)}$$

Our new problem is 49 ÷ 7 which has the answer 7.

Example 5: Divide 324 ÷ 1.8

We write this problem in fraction form $\frac{324}{1.8}$

We multiply both numerator and denominator by 10 to get rid of the decimal point in the divisor. This changes the problem to a whole number division.

$$\frac{324}{1.8} \underset{\times 10}{\overset{\times 10}{=}} \frac{3240}{18} = 180$$

We move the decimal point ONE place to the right in BOTH divisor and dividend.

$$1.8\overline{)324.} = 1.8\overline{)324.0} = 18\overline{)3240} = 180 \quad \text{(quotient } 180\text{)}$$

RULE 2: We can divide any number by a decimal number if we change the *divisor* to a whole number by moving the decimal point the appropriate number of places to the *right.*

We then move the decimal point of the *dividend* the same number of places to the *right*.

We then proceed to divide as in Rule 1.

Example 6: Divide $.027\overline{).810}$

We move the decimal point in the divisor .027 THREE places to the right. We then move the decimal point in the dividend .810, THREE places to the right.

Our problem looks like this:

$$.027\overline{).810} = .027\overline{).810} = 27\overline{)810} = 30 \quad \text{(quotient } 30\text{)}$$

Example 7: 1.05 ÷ .0035

We move the decimal point of both numbers FOUR places to the right, since the divisor .0035 has FOUR places. Our problem then becomes:

$$.0035\overline{)1.05} = .0035\overline{)1.0500} = 35\overline{)10500}\quad 300$$

The answer is 300.

There is one special type of division problem left to consider. It involves dividing a whole number by another whole number.

For example: 5 divided by 8 or $8\overline{)5}$.

We can perform this division if we put a decimal point after the 5 and attach zeros. We then proceed to divide the numbers until we get a zero remainder or a repeating quotient.

Example 8: $5 \div 8$

$5 \div 8$ becomes $8\overline{)5} = 8\overline{)5.000\ldots}$

We keep attaching zeros to 5 after the decimal point, as we need them. We no longer need them when the division stops

$$\begin{array}{r} .625 \\ 8\overline{)5.000} \\ -4\,8 \\ \hline 20 \\ -16 \\ \hline 40 \\ -40 \\ \hline 0 \end{array}$$

We only need 3 zeros since the division stopped after 3 places.

Our answer is .625

Since $5 \div 8$ is also written as $\frac{5}{8}$ the "decimal form" of the fraction $\frac{5}{8}$ is .625.

Example 9: Divide: $5 \div 9$

$5 \div 9$ becomes $9\overline{)5} = 9\overline{)5.000\ldots}$

We keep attaching zeros to 5 as we need them.

We no longer need them when we notice a "repeating" pattern.

$$\begin{array}{r} .555 \\ 9\overline{)5.000} \\ -\underline{4\,5} \\ 50 \\ -\underline{45} \\ 50 \\ -\underline{45} \\ 5 \end{array}$$

No matter how many zeros we attach, the remainder is always "5".

We then get a "repeating decimal" answer.

$\frac{5}{9} = .\overline{5}$ (The bar over the 5 indicates that the digit 5 repeats indefinitely.

Example 10: Divide: $3\overline{)100}$

$3\overline{)100}$ is the same as $100 \div 3$ or $\frac{100}{3}$.

We attach zeros to the 100 after putting the decimal point.

We then proceed to divide continually until we notice a repeating pattern.

$$\begin{array}{r} 33.33 \\ 3\overline{)100.00} \\ -\underline{9} \\ 10 \\ -\underline{9} \\ 10 \\ -\underline{9} \\ 10 \\ -\underline{9} \\ 1 \end{array}$$

No matter how many zeros we attach, the remainder is always 1.

We then get a "repeating decimal" answer.

$3\overline{)100} = 33.\overline{3}$ (The bar over the 3 indicates that the digit 3 continues indefinitely.)

Example 11: Divide: $5 \div 11$

$5 \div 11 = 11\overline{)5} = 11\overline{)5.000\ldots}$

We again attach zeros to the 5 after the decimal point.

```
      .4545
 11)5.0000
   -4 4
      60
    -55
      50
     -44
      (6)0 ── Here, the remainders of 6 and 5 alternate.
     -55   /  This causes the quotient to repeat in the
       (5)    pattern 454545, etc.
```

$5 \div 11 = \frac{5}{11}$ is the repeating decimal $.\overline{45}$.

(The bar over 45 indicates that those digits repeat indefinitely.)

DIVISION OF DECIMALS EXERCISES:

Practice Set A

In each of the following exercises, place the decimal point correctly. The numeric parts of the quotient are correct:

1) $\begin{array}{r} 3 \\ 35\overline{).105} \end{array}$

2) $\begin{array}{r} 6 \\ 18\overline{).108} \end{array}$

3) $\begin{array}{r} 2\ 06 \\ 8\overline{)16.48} \end{array}$

4) $\begin{array}{r} 15 \\ 15\overline{)2.25} \end{array}$

5) $\begin{array}{r} 25 \\ 19\overline{).0475} \end{array}$

6) $\begin{array}{r} 114 \\ 28\overline{).3192} \end{array}$

7) $\begin{array}{r} 36 \\ 105\overline{)37.80} \end{array}$

8) $\begin{array}{r} 1\ 5 \\ 27\overline{)40.5} \end{array}$

9) $\begin{array}{r} 13 \\ 13\overline{).00169} \end{array}$

10) $\begin{array}{r} 4 \\ .25\overline{)1.00} \end{array}$

11) $\begin{array}{r} 22 \\ 1.4\overline{)3.08} \end{array}$

12) $\begin{array}{r} 45 \\ .005\overline{)225} \end{array}$

13) $\begin{array}{r} 52 \\ 3.6\overline{)18.72} \end{array}$

14) $\begin{array}{r} 18 \\ 1.8\overline{).0324} \end{array}$

15) $\begin{array}{r} 3 \\ .0009\overline{)27} \end{array}$

16) $\begin{array}{r} 7 \\ .022\overline{).000154} \end{array}$

17) $\begin{array}{r} 36 \\ 100\overline{)36\quad} \end{array}$

18) $\begin{array}{r} 666 \\ 3\overline{)2\quad\ } \end{array}$

19) $\begin{array}{r} 75 \\ 8\overline{)6\quad} \end{array}$

20) $\begin{array}{r} 133333 \\ 3\overline{)400\quad} \end{array}$

Do the following division problems without a calculator:

21) $7.6 \div 16$

22) $7.2 \div 18$

23) $.192 \div 16$

24) $.0056 \div 7$

25) $11\overline{).00121}$

26) $9\overline{).99}$

27) $25\overline{)1002.5}$

28) $.9\overline{)99}$

29) $1.5\overline{)7.5}$

30) $.003\overline{).693}$

31) $72.48 \div .24$

32) $219.095472 \div 1.304$

33) $.0197\overline{)1.182}$

34) $16 \div 5$

35) $8\overline{)3}$

36) $46\overline{)23}$

37) $3 \div 7$

38) $5 \div 24$

39) $\frac{4}{5}$

40) $3000\overline{).006}$

41) $25\overline{)1}$

42) $178 \div 5$

43) $3\overline{)200}$

44) $11\overline{)100}$

DIVISION OF DECIMALS EXERCISES:

Practice Set B

In each of the following exercises, place the decimal point correctly. The numeric parts of the quotient are correct.

1) $\begin{array}{r} 4 \\ 45\overline{).180} \end{array}$

2) $\begin{array}{r} 9 \\ 12\overline{)1.08} \end{array}$

3) $\begin{array}{r} 8\ 06 \\ 7\overline{)56.42} \end{array}$

4) $\begin{array}{r} 12 \\ 16\overline{)1.92} \end{array}$

5) $\begin{array}{r} 25 \\ 17\overline{).0425} \end{array}$

6) $\begin{array}{r} 152 \\ 38\overline{).5776} \end{array}$

7) $\begin{array}{r} 41 \\ 213\overline{)87.33} \end{array}$

8) $\begin{array}{r} 1\ 75 \\ 34\overline{)59.50} \end{array}$

9) $\begin{array}{r} 14 \\ 14\overline{).00196} \end{array}$

10) $\begin{array}{r} 4 \\ .75\overline{)3.00} \end{array}$

11) $\begin{array}{r} 24 \\ 1.3\overline{)3.12} \end{array}$

12) $\begin{array}{r} 88 \\ .006\overline{)528} \end{array}$

13) $\begin{array}{r} 45 \\ 7.2\overline{).324} \end{array}$

14) $\begin{array}{r} 17 \\ 1.7\overline{).0289} \end{array}$

15) $\begin{array}{r} 7 \\ .0012\overline{)84} \end{array}$

16) $\begin{array}{r} 9 \\ .053\overline{).000477} \end{array}$

17) $\begin{array}{r} 451 \\ 1000\overline{)451\quad} \end{array}$

18) $\begin{array}{l} \quad 66666 \text{ etc.} \\ 6\overline{)4\qquad\quad} \end{array}$

19) $\begin{array}{r} 875 \\ 8\overline{)7\quad} \end{array}$

20) $\begin{array}{r} 3216666 \\ 6\overline{)193\qquad} \end{array}$

Do the following division exercises without a calculator:

21) $8.3 \div 16$

22) $7.6 \div 19$

23) $.288 \div 18$

24) $.00049 \div 7$

25) $19\overline{).0399}$

26) $8\overline{).088}$

27) $.8\overline{)88}$

28) $.45\overline{)9540}$

29) $1.5\overline{)1.08}$

30) $.008\overline{).648}$

31) $28.98 \div .046$

32) $804.6633 \div 2.451$

33) $.0347\overline{).03123}$

34) $5 \div 16$

35) $16\overline{)6}$

36) $92\overline{)23}$

37) $2 \div 7$

38) $5 \div 12$

39) $\frac{6}{5}$

40) $700\overline{).0014}$

41) $30\overline{)9}$

42) $78 \div 4$

43) $8\overline{)97}$

44) $11\overline{)400}$

APPLICATIONS OF MULTIPLICATION AND DIVISION OF DECIMALS EXERCISES:

Practice Set A

1) A train took 5.4 hours to travel 812.7 miles. What was the average speed of the train per hour?

2) If a snail can travel 2.5 centimeters in an hour, how many hours will it take him to cross a 175 centimeter sidewalk? How many days is this? (Round the number of days to 2 decimal places.)

3) A baker can make 130 doughnuts in 3.5 hours. How many hours will he need to complete an order for 325 doughnuts?

4) In a certain chemical solution the ratio of salt to water is 2 to 5. If you have 33.9 milliliters of water, how many grams of salt do you have? (Hint: Use ratio and proportion)

5) A certain box is 177.8 centimeters long. If 1 inch is equivalent to 2.54 centimeters, give the length of the box in INCHES. (Hint: Use ratio and proportion)

6) At a recent school walkathon, a certain local business agreed to give $.46 for every $1.00 that was raised by the event. If the business gave $1150 at the end, how much money did the event make in TOTAL?

7) One pencil weighs 10.25 grams. If the pencils of a certain math class weigh 246 grams, how many pencils are there in the class?

8) Chicken sells for $1.19 per pound. You pick out a bird that weighs 3.45 pounds. How much will this cost you? (Give answer to the nearest cent.)

9) Which is a better buy? 48 pounds of coffee for $312 or 60 pounds of coffee for $384?

10) If a paper clip weighs 1.5 grams on the average, and it costs .975 cents, how much will 1 kilogram (1000 grams) of clips cost? Round your answer to the nearest cent. (Hint: Use ratio and proportion).

11) In most European countries, scales weigh in kilograms instead of pounds. If 2.2 pounds equals 1 kilogram, how many kilograms would a 125 lb woman weigh? (Give answer to 2 decimal places. Use ratio and proportion.)

12) Your bicycle radius, r, is 9 inches. Find its circumference given by the formula $C = 2\pi r$ where $\pi = 3.14$ and $r = 9$.

13) If it takes a computer .0083 seconds to print a digit, how long will it take to print out the number π to 100,000 digits?

14) You are buying soap powder for your laundry. You try not to buy large sizes if you can avoid it. Which is a better buy? 72 ounces of your favorite detergent for $3.99 or 26 ounces for $1.79?

15) A box containing 7 packets of vanilla shake mix, each packet weighing .6 ounces, sells for $2.89. How much per pound is the shake mix costing you? (recall that 1 pound is 16 ounces.) Round your answer to the nearest cent.

16) Your city requires a transfer tax to be paid whenever a house is sold. That tax is $1.10 per $1000 of the selling price. How much tax is paid on a house that sells for $180,000?

17) You calculate the monthly amount to be paid to the bank on your mortgage as $10.29 for each $1000 of the loan. If your loan is for $112,000, what is your monthly mortgage payment?

18) A map of the United States is scaled so that .75 inches represents 100 miles. How many miles is it between San Diego and Seattle if the distance on the map measures 9 inches?

19) In a factory, 4775 machine parts were manufactured. When they were tested, .04 of them were found to be defective. How many parts were defective?

20) When purchased in units of 25, the $1.75 Havana Specials cost $35.00, and the Havana Regulars that usually cost $1.00 each sell for $18.00 per 25. How much would it cost to buy 60 Havana Specials and 40 Havana Regulars at the special rate?

APPLICATIONS OF MULTIPLICATION AND DIVISION OF DECIMALS EXERCISES

Practice Set B

1) The British gallon of gasoline is equivalent is to 1.20094 United States gallons. If an automobile travels 32 miles on one British gallon of gasoline, how many United States gallons would it use to travel 288 miles? (Round answer to 1 decimal place. Use ratio and proportion.)

2) The value of the dollar as compared to the British pound fluctuates day by day. Suppose that 1 British pound can be bought for $1.52 US dollars. You are in England and you want to pay for an item that costs 25 pounds. How much is that equivalent to in US dollars?

3) Each dime coin weighs .0022 pounds. How many dimes would there be in stack that weighs just under 1 pound?

4) Which is a better buy? A 15.5 ounce can of water packed tuna for $1.79 or a 6.5 ounce can for $.76?

5) The largest tomato ever grown was in Wisconsin in 1976 and weighed 6.5 pounds. If a recipe for tomato sauce calls for .75 pounds of fresh tomato, and serves 4 people, how many people could have been served by this giant tomato?

6) You choose a flank steak from your butcher that weighs 4.27 pounds. It is on special this week for $4.09 per pound. The butcher forgets to change the price on the scale and it reads $21.31 instead of what it should. How much should you have been charged? How much per pound were you overcharged?

7) Thanksgiving turkeys always sell for a reasonable price since there are so many of them bred for that time of year. If you choose a bird that weighs 24.0785 pounds on the scale and turkey sells for $1.18 per pound, how much will that turkey cost? (round off answer to the nearest cent.)

8) A local natural foods shoppe sells spices by the ounce. You are making your own curry powder which uses a mixture of turmeric and cumin powders. The turmeric powder sells for \$.98 per ounce while the cumin sells for \$1.17 per ounce. What is the cost of a one pound mixture that contains 10 ounces of cumin and the rest turmeric?

9) The recent world birth rate was approximated as 27.3 births per 1000 population. The death rate was calculated as 10.6 per 1000 population. If the world's total population is 4,865,000,000 people,

a) How may babies are born each year?

b) How many people die each year?

CHANGING FRACTIONS TO DECIMALS AND DECIMALS TO FRACTIONS:

Any fraction can be thought of as representing a division problem. We can use our techniques of division to arrive at the decimal form for any fraction. We simply change the fraction to a division problem and proceed to divide. We continue dividing until we have a zero remainder or a repeating set of remainders.

Example 1: Change $\frac{3}{8}$ to decimal form.

We simply change the $\frac{3}{8}$ to the division problem $8\overline{)3.00}$. We then do the division.

$$\begin{array}{r}
.375 \\
8\overline{)3.000} \\
-\underline{24} \\
60 \\
-\underline{56} \\
40 \\
-\underline{40} \\
0
\end{array}$$

So $\frac{3}{8} = .375$

Example 2: Change $\frac{2}{3}$ to decimal form.

We first change $\frac{2}{3}$ to the division problem $3\overline{)2.000}$

In doing this division, the remainder 2 will repeat indefinitely.

$$\begin{array}{r}
.666 \\
3\overline{)2.000} \\
-\underline{1\,8} \\
20 \\
-\underline{18} \\
20 \\
-\underline{18} \\
2
\end{array}$$

We have what is called a *"repeating decimal"*. We indicate this by putting a bar over the portion of the quotient that repeats.

$\frac{2}{3} = .\overline{6}$ (The bar is over the 6 since that digit repeats)

Example 3: Change $\frac{5}{6}$ to decimal form.

We do the problem

$$\begin{array}{r} .8333 \\ 6\overline{)5.0000} \\ -4\,8 \\ \hline 20 \\ -18 \\ \hline 20 \\ -18 \\ \hline 20 \\ -18 \\ \hline 2 \end{array}$$

We continually get a remainder of 2 which gives us a 20 to divide by 6. The answer will have a string of 3s at the end for this reason.

We write $\frac{5}{6} = .8\overline{3}$ (with the bar just over the 3 since it is just the 3s that repeat indefinitely.)

Example 4: Change $\frac{7}{11}$ to decimal form.

We change $\frac{7}{11}$ to the division problem $11\overline{)7.000}$

$$\begin{array}{r} .6363 \\ 11\overline{)7.0000} \\ -6\,6 \\ \hline 40 \\ -33 \\ \hline 70 \\ -66 \\ \hline 40 \\ -33 \\ \hline 7 \end{array}$$

The division continually generates remainders of 4 and 7. This gives us a string of 63s in the quotient. Thus, $\frac{7}{11}$ is written in decimal form as $.\overline{63}$. (The bar is over the 63 since that pair of digits repeats indefinitely.)

$$\frac{7}{11} = .\overline{63}$$

Changing decimals to fractions is a bit more involved. If the decimal is finite (the digits don't repeat indefinitely) then expressing the decimal as a fraction is simply a matter of writing the decimal as it reads. Then it is a fraction with a denominator that is a power of 10.

Example 5: Change .875 to fraction form.

.875 is read as "875-thousandths" $= \frac{875}{1000}$

We can then reduce the fraction to lowest terms.

$$\frac{875}{1000} \overset{\div 125}{=} \frac{7}{8} \quad (\div 125 \text{ top and bottom})$$

Example 6: Change 25.3 to fraction form.

25.3 is read as "twenty five and 3-tenths"

which is $25\frac{3}{10} = \frac{253}{10}$.

Changing *repeating* decimals to fraction form is more complicated. It involves algebraic reasoning. I will demonstrate some sample problems. See if you can follow the pattern.

Example 7: Change the repeating decimal .3 to fraction form.

$.\overline{3} = .3333333$ (3s go indefinitely). Call the number "x".

x = .3333333 etc.

We will multiply "x" by 10. Call that number "10x".

10x = 10(.3333333 etc.) = .3 3333333 etc. = 3.3333333 etc.

10x = 3.3333333 etc.

$$\begin{array}{rl} 10x = & 3.3333333 \text{ etc.} \\ -\quad x = & .3333333 \text{ etc.} \\ \hline 9x = & 3 \end{array}$$

We subtract one "x" from 10x and get 9x.

(When subtracting, the repeating parts of the decimals are the same, so they subtract out to zero.)

Just as we did in our section on Ratio and Proportion, we *divide* by 9 to get "x".

9x = 3

$x = \frac{3}{9} = \frac{1}{3}$

Therefore $.\overline{3} = \frac{1}{3}$

Example 8: Change $.1\overline{6}$ to fraction form.

$.1\overline{6} = .16666666$ etc. Call it "x".

x = .1666666 etc.

Since the number repeats from the *second* decimal place, we multiply the number by 100. We call that number 100x.

100x = 100(.1666666 etc.) = . 1 6 6 6 6 6 6 etc. = 16.66666 etc.

100x = 16.66666 etc.

We also get a number "10x" by multiplying by 10.

10x = 1.66666 etc.

We then proceed to subtract the two numbers. This can be done since the repeating decimal parts are the same.

$$\begin{array}{rrl} & 100x = & 16.66666 \text{ etc.} \\ - & 10x = & 1.66666 \text{ etc.} \\ \hline & 90x = & 16 - 1 = 15 \\ & 90x = & 15 \end{array}$$

$$x = \frac{15}{90} = \frac{1}{6} \quad (\div 15) \qquad \text{Therefore } .1\overline{6} = \frac{1}{6}$$

Example 9: Change $.\overline{45}$ to fraction form.

$.\overline{45} = .45454545$ etc. Call the number "x".

We multiply the number by 100. We call it "100x".

100x = 45.45454545 etc.

We can then subtract the number "x" from "100x" since the repeating parts of the numbers are the same.

$$\begin{array}{rrl} & 100x = & 45.45454545 \text{ etc.} \\ - & x = & .45454545 \text{ etc.} \\ \hline & 99x = & 45 \end{array}$$

$$x = \frac{45}{99} = \frac{5}{11} \quad (\div 9) \qquad \text{Therefore: } .\overline{45} = \frac{5}{11}.$$

CHANGING FRACTIONS TO DECIMALS AND DECIMALS TO FRACTIONS EXERCISES

Practice Set A

Put the following fractions in decimal form. Carry out all divisions until the remainder is zero or repeats in a pattern.

1) $\frac{3}{5}$	2) $\frac{1}{9}$	3) $\frac{19}{50}$	4) $\frac{5}{3}$	5) $\frac{21}{10}$
6) $\frac{7}{8}$	7) $\frac{5}{12}$	8) $\frac{2}{11}$	9) $\frac{7}{6}$	10) $\frac{12}{5}$
11) $\frac{200}{3}$	12) $\frac{25}{2}$	13) $\frac{1}{4}$	14) $\frac{3}{4}$	15) $\frac{999}{1000}$
16) $\frac{1}{6}$	17) $\frac{1}{25}$	18) $\frac{78}{33}$	19) $\frac{100}{16}$	20) $\frac{1}{20}$

21) There is an interesting pattern to discover in the decimal forms of the fractions with denominator 7.

a) Find the decimal forms of each of the following sevenths:

$\frac{1}{7} =$

$\frac{2}{7} =$

$\frac{3}{7} =$

$\frac{4}{7} =$

$\frac{5}{7} =$

$\frac{6}{7} =$

b) What do all the decimal forms of the sevenths have in common?

c) How can you use the decimal form of $\frac{1}{7}$ to get the other decimal forms? What is the pattern between the decimal forms?

22) a) Change $\frac{1}{11}$ and $\frac{1}{111}$ to decimal form.

b) What is the relation between the two fractions and their decimal forms?

c) Change $\frac{2}{11}$ and $\frac{2}{111}$ to decimal form.

d) What is the relation between the two fractions in part (c) and their decimal forms?

e) Find the decimal form for $\frac{5}{11}$.

f) Predict the decimal form for $\frac{5}{111}$.

g) Predict the decimal form for $\frac{5}{1111}$.

Change the following decimals to fractions in most reduced form. You can do this by reading off the decimals.

23) .75 24) .125 25) .805 26) .8 27) .4

28) 7.25 29) 10.35 30) 2.0625 31) 1.06 32) 15.2

33) 100.05

Change the following decimals to fraction form by using our method of "algebraic reasoning." These are difficult.

34) $.\overline{6}$ 35) $.8\overline{3}$ 36) $.\overline{2}$ 37) $.\overline{75}$ 38) $.12\overline{3}$

39) $.1\overline{23}$ 40) $.\overline{123}$

CHANGING FRACTIONS TO DECIMALS AND DECIMALS TO FRACTIONS EXERCISES

Practice Set B

Put the following fractions in decimal form:

1) $\frac{4}{5}$ 2) $\frac{2}{9}$ 3) $\frac{17}{25}$ 4) $\frac{7}{3}$ 5) $\frac{63}{10}$

6) $\frac{3}{8}$ 7) $\frac{7}{12}$ 8) $\frac{3}{11}$ 9) $\frac{7}{5}$ 10) $\frac{19}{20}$

11) $\frac{100}{3}$ 12) $\frac{37}{4}$ 13) $\frac{5}{8}$ 14) $\frac{9}{16}$ 15) $\frac{789}{1000}$

16) $\frac{23}{64}$ 17) $\frac{3}{25}$ 18) $\frac{95}{66}$ 19) $\frac{200}{12}$ 20) $\frac{3}{40}$

21) There is an interesting pattern to discover in the decimal forms of the fractions with denominator 11.

a) Find the decimal forms of each of the following elevenths:

$\frac{1}{11} =$

$\frac{2}{11} =$

$\frac{3}{11} =$

$\frac{4}{11} =$

$\frac{5}{11} =$

$\frac{6}{11} =$

$\frac{7}{11} =$

$\frac{8}{11} =$

$\frac{9}{11} =$

$\frac{10}{11} =$

b) What is the pattern that predicts the decimal form for any elevenths?

c) Predict the decimal forms of $\frac{12}{11}$, $\frac{13}{11}$, $\frac{14}{11}$ and $\frac{15}{11}$.

Change the following decimals to fractions in most reduced form.

22) .075 23) .625 24) .545 25) .9 26) .33

27) 11.25 28) 100.6 29) 4.0075 30) 2.40 31) 18.6

32) 99.005

Change the following decimals to fraction form by using our method of "algebraic reasoning." These are difficult.

33) $.\overline{4}$ 34) $1.1\overline{6}$ 35) $.\overline{7}$ 36) $.\overline{25}$ 38) $.54\overline{2}$

38) $.5\overline{42}$ 39) $.\overline{542}$

DECIMALS AND THE CALCULATOR

The calculator was designed to do all decimal problems. Fractions require more thought, but decimals can be entered immediately into the calculator. We can do any arithmetic problem with decimals. We can even change fractions to decimals. The difficulty is when we want to change decimals to fractions. Calculators are not set up to give fractional answers, only decimal answers. Here are some various problems worked on the calculator:

Example 1: Add 12 + .095 + .00078

We simply enter the numbers in the calculator and add.

We would do this: 12 (+) .095 (+) .00078 (=)

The answer would appear in the display as 12.09578

Example 2: Subtract 98.6 − 9.875

We simply enter the problem as it reads.

The steps are 98.6 (−) 9.875 (=)

The answer would appear as 88.725

Example 3: Multiply 97.6 × 89.9

We simply enter the problem as it reads.

The steps are 97.6 (×) 89.9 (=)

The answer appears on the display as 8774.24

Example 4: $1.2\overline{).0468}$

The *dividend* is always entered *first*, followed by the operation ÷, and then the divisor.

The steps are: .0468 (÷) 1.2 (=)

The answer will appear as .039

Example 5: Change the fraction $\frac{3}{8}$ to decimal form.

We enter this problem as the division problem 3 ÷ 8.

The steps are 3 (÷) 8 (=)

The answer is .375

Example 6: Change the fraction $\frac{2}{3}$ to decimal form.

We enter the fraction as the division problem 2 (÷) 3 (=).

In the display we see the number .666666667

The calculator *rounds off* the number after a certain number of digits. The number of digits depends on your particular calculator.

DECIMALS — REVIEW EXERCISES

Practice Set A

Write the decimal that goes with each written expression:

1) 15 and 23-thousandths

2) 101-ten thousandths

3) 2456 and 39-hundredths

4) 2 and 2-millionths

Write out how you would read the following numbers using the correct place values:

5) 2.018

6) 1987.0045

7) 6.00006

Change each of the following decimals to FRACTION form. Reduce answers wherever possible.

8) 8.625

9) 14.025

10) $37.\overline{33333}$ (a string of 3s)

11) 298.9

12) 1987.0075

Order the following decimals from smallest to largest:

13) 2.8, 2.88, 2.8801, 2.0888, 2.089, 2.88009

14) 1.75, 1.749, 1.7501, 1.755, 1.75009

15) 1.0001, 1.001, 1.00011

Add or subtract as the problem dictates. Do not use a calculator. It helps to set the problems up vertically.

16) 7.56 + .003 + .1574

17)
```
  99.891
  12
   2.0004
+ 101.1
```

18) 99.2 − 78.98

19) 80 − 35.7659

20) 75 − (.0075 + 2.4)

Write each of the following numbers in scientific notation:

21) 245,000,000 22) 17.98 23) 800

24) 40,440,000,000,000 25) 9.23 26) 10 trillion

In each of the following exercises, put the decimal point in the correct place. The numeric parts of the answer are correct.

27) $2.28 \times .03 = 6\,8\,4$ 28) $29.4 \times 1.3 = 3\,8\,2\,2$

29) $.9 \times .3 \times 1.1 = 2\,9\,7$ 30) $7500 \times .025 = 1\,8\,7\,5$

31) $\begin{array}{r} 4 \\ 29\overline{)1.16} \end{array}$ 32) $\begin{array}{r} 18 \\ 17\overline{).306} \end{array}$ 33) $\begin{array}{r} 17 \\ .005\overline{).85} \end{array}$

34) $\begin{array}{r} 11 \\ .043\overline{).473} \end{array}$ 35) $\begin{array}{l} \quad 2\,3\,3\,3\,3\,3\,3\,3 \\ 3\overline{)7\qquad\qquad\quad} \end{array}$

Multiply or divide as indicated in the following exercises. Do not use a calculator.

36) $.45 \times 4$ 37) $.19 \times 1.8$ 38) $.05 \times .9 \times .3$

39) $(2.4)^2$ 40) $7.8 \times .0006 \times .2$ 41) $.00049 \div 7$

42) $5.85 \div 15$ 43) $.006\overline{)1.2183}$ 44) $19 \div 5$

45) $2 \div 9$

Put the following fractions in decimal form. Carry out all divisions until there is a zero remainder or the remainder repeats in a pattern:

46) $\frac{17}{8}$ 47) $\frac{9}{4}$ 48) $\frac{7}{12}$ 49) $\frac{13}{20}$ 50) $\frac{5}{24}$

Change the following decimals to fractions in most reduced form.

51) .745 52) 13.0035 53) 2.875

54) 19.04 55) $.\overline{7}$ 56) $.5\overline{6}$

57) A man has 289.75 feet of fencing. Would he have enough to enclose this field? If not, how much would he need? If so, does he have any fencing left over?

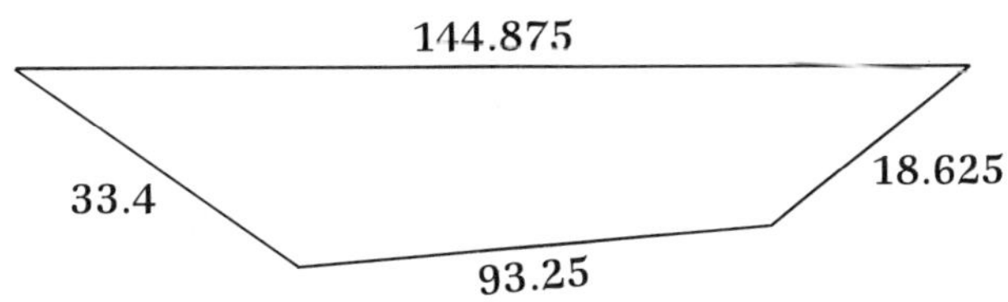

58) In a certain chemical solution, the ratio of hydrochloric acid to water is 3 to 4. If there are 63.9 milliliters of hydrochloric acid, how many milliliters of water are there?

59) Suppose that each French franc is worth $.1423 US dollars. How many French francs can be bought for $71.15?

60) If 3.75 pounds of coffee cost $22.50, find the cost of 9 pounds of coffee.

DECIMALS — REVIEW EXERCISES

Practice Set B

Write the decimal that goes with each written expression:

1) 27 and 35-thousandths

2) 206-ten thousandths

3) 5431 and 45-hundredths

4) 3 and 3-ten thousandths

Write out how you would read each of the following numbers using the correct place values:

5) 7.028

6) 1987.0036

7) 8.00008

Change each of the following decimals to FRACTION form. Reduce answers wherever possible.

8) 7.645

9) 19.004

10) $98.\overline{66666}$ (a string of 6s)

11) 453.8

12) 2487.0325

Order the following decimals from smallest to largest:

13) 7.45, 7.44956, 7.49901, 7.499, 7.5

14) .835, .83489, .8349, .836, .84

Add or subtract as the problem dictates. Do not use a calculator. It helps to set the problems up vertically.

15) 2.13 + .007 + .02735

16)
$$\begin{array}{r} 185 \\ 78.005 \\ 2.1 \\ +\quad .4751 \\ \hline \end{array}$$

17) 1000 − 345.9874

18) 112 − 95.0045

19) 28 − (15.375 − 4.98)

Write each of the following numbers in scientific notation:

20) 3,058,000,000,000　　21) 7.5　　22) 3000

23) 9,006,421,000,000,000　　24) 8.71　　25) 10 trillion, 100 million

In each of the following exercises, put the decimal point in the correct place. The numeric parts of the answer are correct.

26) $7.41 \times .009 = 6\,6\,6\,9$　　27) $5.86 \times 2.4 = 1\,4\,0\,6\,4$

28) $.12 \times .0035 \times .4 = 1\,6\,8\,0$　　29) $2950 \times .005 = 1\,4\,7\,5\,0$

30) $\begin{array}{r} 18 \\ 26\overline{).468} \end{array}$　　31) $\begin{array}{r} 95 \\ 13\overline{).01235} \end{array}$　　32) $\begin{array}{r} 15 \\ .0075\overline{).1125} \end{array}$

33) $\begin{array}{r} 5\;99 \\ .4\overline{)23.96} \end{array}$　　34) $\begin{array}{l} \quad 366666 \\ 3\overline{)11\qquad\;} \end{array}$　　35) $\begin{array}{l} \qquad 916666 \\ 12\overline{)11\qquad\quad\;} \end{array}$

Multiply or divide as indicated in the following exercises. Do not use a calculator.

36) $.83 \times 12$　　37) $.015 \times 1.6$　　38) $.003 \times .14 \times .5$

39) $(4.5)^2$　　40) $9.2 \times .0004 \times .3$　　41) $.00084 \div 1.2$

42) $7 \div 2.5$　　43) $18 \div 5$　　44) $3\overline{)11\;}$

Put the following fractions in decimal form. Carry out all divisions until there is a zero remainder or the remainder repeats in a pattern:

45) $\frac{23}{8}$　　46) $\frac{17}{4}$　　47) $\frac{11}{24}$

48) $\frac{15}{40}$　　49) $\frac{3}{16}$　　50) $\frac{9}{25}$

Change the following decimals to fractions in most reduced form.

51) .825　　52) 23.1875　　53) 7.625　　54) 4.0025　　55) $.\overline{8}$　　56) $.4\overline{5}$

57) Rauncy Ray, the sly antique dealer, sold a certain antique watch for $456.78, bought it back for $473.23 and then resold it for $515.42. What did he make on the transaction?

58) If 35 packages cost $48.19, find the cost of 245 such packages.

59) Sean buys a new sports car to drive in. It has a stick shift. Six months later, Sean breaks his leg and he must be in a cast for 1 year. He decides to sell the car. He sells it for .2 less than what he paid for it. His loss is $2900. How much did the sports car cost?

60) A plane flies 700 miles in 4.375 hours. What is its average speed per hour?

DECIMALS — REVIEW OF IMPORTANT WORDS

Here is a checklist of words that are used in dealing with decimals. See if you can identify each one and define it to yourself.

decimal point

place value (positional value)

fraction form

decimal form

equivalent decimal numbers

expanded form

powers of 10

moving a decimal point a certain number of places (right or left)

(over)

repeating decimal

finite decimal

SAMPLE DECIMALS EXAM

Show all work. Do not use calculator.

1) 1856.003 + 44.5 + 8 + .21063

2) .018 + 18.15 + 143 + .21007

3) 8.47 − 3.29

4) 500 − 6.983

5) $\begin{array}{r} 18.15 \\ \times\ \ 7.2 \\ \hline \end{array}$

6) $\begin{array}{r} 21.32 \\ \times\ \ 5.5 \\ \hline \end{array}$

7) $216\overline{)72}$

8) $47\overline{).564}$

9) $1.9\overline{)4.75}$

10) $2.8\overline{)2.24}$

11) Place the decimal points correctly:

a) .0047 × .0096 = 4 5 1 2

b) 2.95 × 3.14 = 9 2 6 3 0

12) Change these fractions to decimals:

a) $\frac{19}{100}$ b) $\frac{7}{8}$ c) $\frac{18}{27}$

13) Write these decimals as fractions (reduce if possible):

a) .375 b) .025 c) .0015

14) Arrange these in order, largest to smallest: .65, .661, .649, .649001

15) A man had 752.25 feet of fencing. How close could he come to being able to enclose this field?

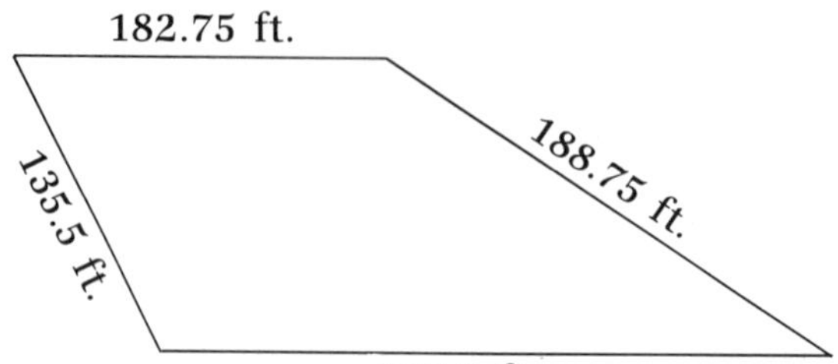

16) Perform these operations:

a) 87.6×1000

b) $.285 \div 100$

c) $.009 \times 10$

d) $.6 \div 10$

17) Use your calculator to compute these:

a) $953.8 \times .017$

b) $.00375 \times 956.23$

c) $2195.125 \div .05$

d) $.48\overline{)5.952}$

e) $\frac{15}{11} =$

f) $2.7 - .27 - (.2 + .007)$

g) $(2.36)^2$

h) $.8 \div 10 \div .01$

i) $7.54 \times 2.36 \times 1000$

j) $\frac{5}{16} =$

k) $6 + 60 + .6 + .06 + .006$

l) $\$1,234,576.42 - \$1.19$

m) Change to decimal: $2\frac{17}{24}$

"EXCUSE ME, BUT DID I SAY A 1% OR A 100% SOLUTION OF DR. GROWRITE'S HORMONE ELIXER THIS TIME?"

CHAPTER 4: PERCENT

MEANING AND DECIMAL FORM

We can think of the word "percent" as a literal translation of the Latin "per centum" which means "per hundred".

Whenever we see the word "percent" or the sign "%", we can replace it with the word "hundredth" or the symbol "$\frac{1}{100}$".

PERCENT means *HUNDREDTHS*.

% = hundredths

Example 1: 12% means 12-hundredths $= \frac{12}{100}$

Example 2: 250% means 250-hundredths $= \frac{250}{100}$

Example 3: $7\frac{1}{2}\%$ means $7\frac{1}{2}$ hundredths = 7.5-hundredths $= \frac{7.5}{100}$

Example 4: .25% means .25-hundredths $= \frac{.25}{100}$

Once we have the percent expressed as a fraction with denominator 100, we can easily change that fraction to a decimal. It can be thought of as a division problem, or depending on the problem, converted to a decimal by inspection.

Example 5: $12\% = \frac{12}{100} = .12$ (by inspection)

Example 6: $7\frac{1}{2}\% = \frac{7.5}{100} = 7.5 \div 100 = 07.5 = .075$

Example 7: $250\% = \frac{250}{100} = 250 \div 100 = 250 = 2.5$

Example 8: $.25\% = \frac{.25}{100} = .25 \div 100 = 00.25 = .0025$

There is a shortcut to changing percents to decimals. Looking at the examples just completed, you can see that the answer can be gotten by moving the decimal point *to the left two places.*

Example 9: $12\% = \frac{12}{100} = .12$

We can think of the decimal point in 12, then move it two places to the *left*.

$12\% = \frac{12}{100} = 12 \div 100 = 12. = .12$

Therefore: $12\% = .12$

RULE: PERCENT TO DECIMAL FORM: To change a percent to a decimal, we move the decimal point two places to the *LEFT.*

Example 10: Change 24% to a decimal.

$24\% = 24. = .24$

Example 11: Change 8% to a decimal.

$8\% = 08. = .08$

Example 12: Change $11\frac{1}{2}\%$ to a decimal.

$11\frac{1}{2}\% = 11.5\% = 11.5 = .115$

Many times we are given decimal numbers that need to be changed into percents. We can do this if we *change a decimal to a fraction with denominator 100.* Since percent literally means hundredths, we can replace the denominator of 100 with the % sign after the numerator. (Remember: % = hundredths)

Example 13: Change .75 to a percent.

$$.75 = \frac{75}{100} = \text{75-hundredths} = 75\%$$

We simply replace hundredths by the % sign.

Example 14: Change .089 to a percent.

$.089 = \frac{89}{1000}$ Now we will change $\frac{89}{1000}$ to hundredths.

$$\frac{89}{1000} \underset{\div 10}{\overset{\div 10}{=}} \frac{8.9}{100} = \underbrace{\text{8.9-hundredths} = 8.9\%}_{\text{equal}}$$

Example 15: Change 1.33 to a percent.

$$1.33 = 1\frac{33}{100} = \frac{133}{100} = \text{133-hundredths} = 133\%$$

Example 16: Change .4 to a percent.

$$.4 = \frac{4}{10} \underset{\times 10}{\overset{\times 10}{=}} \frac{40}{100} = \text{40-hundredths} = 40\%$$

In the examples given, the effect of changing a decimal to a percent was to move the decimal point in the number *two* places to the *right*, and then attach a % sign. This shortcut will work every time.

> RULE: DECIMAL TO PERCENT: To change a decimal to a percent, we move the decimal point two places to the *right* and attach a % sign.

Example 17: Change .28 to a percent.

$$.28 = .28 = 28\%$$

Example 18: Change .015 to a percent.

$$.015 = .0\,1\,5 = 1.5\% \text{ or } 1\tfrac{1}{2}\%.$$

Example 19: Change 2.75 to a percent

$$2.75 = 2\,.\,7\,5 = 275\%$$

It is often helpful to *memorize* the common percent forms that we encounter in many applications. Here is a list of useful percents and both their decimal and fraction forms.

$25\% = .25 = \frac{1}{4}$

$50\% = .50 = \frac{1}{2}$

$75\% = .75 = \frac{3}{4}$

$100\% = 1.00 = 1$

$33\frac{1}{3}\% = \overline{.3333} = \frac{1}{3}$

$66\frac{2}{3}\% = \overline{.6666} = \frac{2}{3}$

$12\frac{1}{2}\% = 12.5\% = .125 = \frac{1}{8}$

$37\frac{1}{2}\% = 37.5\% = .375 = \frac{3}{8}$

$62\frac{1}{2}\% = 62.5\% = .625 = \frac{5}{8}$

$87\frac{1}{2}\% = 87.5\% = .875 = \frac{7}{8}$

$20\% = .20 = \frac{1}{5}$

$40\% = .40 = \frac{2}{5}$

$60\% = .60 = \frac{3}{5}$

$80\% = .80 = \frac{4}{5}$

PERCENTS — MEANING AND DECIMAL FORM EXERCISES:

Practice Set A

Change the following percents to decimal form:

1) 19%	2) 35%	3) 76%	4) 9%
5) 45%	6) 215%	7) 132%	8) 450%
9) 155%	10) 98%	11) .25%	12) 12.5%
13) 77.25%	14) 91.6%	15) .3%	16) $17\frac{1}{2}\%$
17) $49\frac{3}{8}\%$	18) $65\frac{1}{4}\%$	19) $6\frac{1}{2}\%$	20) $\frac{3}{4}\%$

Change each of the following decimals to percents:

21) .63	22) .52	23) .07	24) .16
25) .48	26) .93	27) .80	28) .5
29) .03	30) .0025	31) .075	32) 2.04
33) 2.25	34) .028	35) $.\overline{333}$	36) .0125
37) .1875	38) $.06\frac{2}{3}$	39) $.15\frac{1}{2}$	40) 3

PERCENTS — MEANING AND DECIMAL FORM EXERCISES

Practice Set B

Change the following percents to decimal form:

1) 75%
2) 100%
3) .5%
4) 50%
5) 345%
6) 87%
7) 29%
8) 135%
9) 756%
10) 89%
11) .83%
12) $2\frac{1}{2}\%$
13) 17.75%
14) 200%
15) .06%
16) 1.75%
17) $66\frac{2}{3}\%$
18) $37\frac{4}{5}\%$
19) $8\frac{1}{2}\%$
20) .5%

Change each of the following decimals to percents:

21) .68
22) 1.53
23) 1.25
24) .125
25) .0125
26) .78
27) 4
28) $.13\frac{1}{2}$
29) .0475
30) 1.3125
31) .285
32) .09
33) .0075
34) .0001
35) .6666
36) .3425
37) $.08\frac{2}{3}$
38) $.54\frac{1}{2}$
39) 1.5
40) .004

USE OF PERCENTS

Percents are only used to refer to a "*base*" quantity. Percents, as numbers, don't stand alone. We always talk about the "percent *of*" some quantity. *Percent is a way of comparing some part to a whole.*

Example 1: 4 is 50% *of* 8.

We are comparing the 4 to the 8. We get $\frac{4}{8}$.

$\frac{4}{8} = \frac{1}{2} = .5 = 50\%$

Example 2: 18 is 200% *of* 9.

We are comparing the 18 to the 9 or $\frac{18}{9}$.

$\frac{18}{9} = 2 = 200\%$.

The number after the word "*of*" is called the BASE.

Let's call the number associated with the word "*is*" the AMOUNT.

AMOUNT is actually the comparison value; what we are comparing to the BASE.

Any percent is the ratio of AMOUNT to BASE, i.e., $\boxed{\frac{\text{AMOUNT}}{\text{BASE}} = \%}$

A percent sentence can be put in this form:

%		BASE		AMOUNT
______	of	________	is	____________

or

%		BASE		AMOUNT
______	×	________	=	____________

The *ratio* form for percents is $\boxed{\frac{\text{AMOUNT}}{\text{BASE}} = \%}$. This form is easiest to use.

Example 3: What percent *of* 40 is 25?

The BASE is 40 (since it appears after the "of").

The AMOUNT is 25 (since it appears after the "is").

We can set up the ratio form for percents.

$$\frac{\text{AMOUNT}}{\text{BASE}} = \frac{25}{40} = \frac{5}{8} = .625 = 62.5\%$$

The answer is 62.5%.

Example 4: What number is 75% of 80?

The BASE number is 80.

We don't know the AMOUNT. Call it "n".

Let's set up the ratio form for percents.

$$\frac{\text{AMOUNT}}{\text{BASE}} = \frac{n}{80} = 75\% \text{ or } .75$$

$$\frac{n}{80} = .75 \quad \text{or} \quad \frac{n}{80} = \frac{.75}{1}$$

We can find "n" if we use a technique from the "Ratio and Proportion" section.

Recall that "CROSS PRODUCTS ARE EQUAL."

$$\frac{n}{80} \times \frac{.75}{1} \quad \text{becomes } n = .75(80) = 60$$

The answer is n = 60.

The AMOUNT is 60.

Example 5: 10 is 30% of what number?

Here we know the AMOUNT is 10, but we don't know the BASE. Call the BASE number "n".

Setting up the ratio form for percents we get:

$$\frac{\text{AMOUNT}}{\text{BASE}} \; \frac{10}{n} = 30\% \text{ (or .30)}$$

Now we cross multiply to find "n".

$$\frac{10}{n} \times \frac{.30}{1} \quad \text{gives } .30n = 10(1) = 10$$

From .30n = 10 we divide both sides by .30 to get "n" which gives

$$n = \frac{10}{.30} = \frac{100}{3} = 33.\overline{3}$$

The BASE is $33.\overline{3}$

From the exercises just done we can see that all percent exercises break down to three different categories:

I) when PERCENT is unknown
II) when AMOUNT is unknown
III) when BASE is unknown.

Let's see how to approach each type.

TYPE I) When PERCENT is unknown:

Example 6: What percent of 80 is 30?

We set up the ratio form for percents. It will immediately give us the answer.

$$\frac{\text{AMOUNT}}{\text{BASE}} \quad \frac{30}{80} = \%$$

$$\frac{30}{80} = .375 = 37.5\%$$

Example 7: 25 is what percent of 75?

Again, the ratio form for percents immediately gives us the answer.

$$\frac{\text{AMOUNT}}{\text{BASE}} = \frac{25}{75} = \%$$

$$\frac{25}{75} = \frac{1}{3} = \overline{.333} = 33\frac{1}{3}\%$$

(by memory)

TYPE II) When AMOUNT is unknown:

There is a shortcut for doing this type of exercise.

Remember that the word *"of"* means *"multiply"*.

We simply translate the exercise replacing "of" by a "×" sign.

WE MUST CHANGE PERCENTS TO DECIMAL FORM BEFORE USING THEM IN CALCULATIONS.

Example 8: What is 45% of 75?

We translate this to be 45% × 75 or .45 × 75 = 33.75

The answer is 33.75

Example 9: What is 250% of 14?

We translate this to 250% × 14 or 2.50 × 14 = 35.

The amount is 35.

Example 10: What is $66\frac{2}{3}\%$ of 24?

This exercise is easier if we remember that $66\frac{2}{3}\%$ is $\frac{2}{3}$ and use the fraction instead of decimal form.

$$66\frac{2}{3}\% \times 24 = \frac{2}{3} \times \frac{24}{1} = \frac{2}{\cancel{3}_1} \times \frac{\cancel{24}^{8}}{1} = 16.$$

The answer is 16.

TYPE III) when BASE is unknown:

This is the most complicated type of example. It requires an extra step at the end (dividing by the percentage). We must change percents to decimal form in calculations:

Example 11: 13 is 25% of what number?

BASE is unknown so we call it "n".

Here we use the sentence form of percents to get:

25% of BASE = AMOUNT

25% of n = 13

which becomes

.25n = 13

To find "n" we divide by the percent (.25)

$$n = \frac{13}{.25} = 52$$

The answer is 52.

Example 12: 100 is 125% of what number?

The BASE is unknown. Call it "n".

Using the sentence form of percents we get:

125% of n = 100

which becomes

1.25n = 100

To find "n" we divide by 1.25 (the percent in decimal form).

$$n = \frac{100}{1.25} = 80$$

The answer is 80.

Example 13: 15 is $33\frac{1}{3}$% of what number?

Again, call the BASE number "n"

Using the sentence form we get:

$33\frac{1}{3}$% of n = 15

Here it is helpful to change $33\frac{1}{3}\%$ to the *fraction* $\frac{1}{3}$.

We get: $\frac{1}{3}n = 15$

We divide by $\frac{1}{3}$ to get $n = \frac{15}{\frac{1}{3}} = \frac{15}{1} \times \frac{3}{1} = 45$

The answer is 45.

USE OF PERCENTS EXERCISES

Practice Set A

1) What is 25% of 88?

2) What is 150% of $17.00?

3) What is $12\frac{1}{2}$% of 300?

4) What percent of 75 is 15?

5) What percent of 10 is 60?

6) What percent of 50 is .25?

7) What number is 40% of 11?

8) 18 is $2\frac{1}{4}$% of what number?

9) 12 is 40% of what number?

10) 1 is 250% of what number?

11) What is $\frac{3}{4}$% of 400?

12) What number is $13\frac{2}{3}$% of 3000?

13) 678 is what percent of 678?

14) 37 is what percent of 100?

15) 14 is $66\frac{2}{3}$% of what number?

16) 100 is what percent of 125?

17) 125 is what percent of 100?

18) .4 is $\frac{1}{2}$% of what number?

19) What is $66\frac{2}{3}$% of 42?

20) What percent of 72 is 12?

21) 9 is what percent of 24?

22) What is $6\frac{1}{2}$% of $10.00?

23) 1.5 is what percent of .375?

24) 45 is $33\frac{1}{3}$% of what number?

25) What is 350% of 12?

USE OF PERCENTS EXERCISES

Practice Set B

1) What is 60% of 500?

2) What is 125% of $18.00?

3) What is $37\frac{1}{2}$% of 240?

4) What percent of 75 is 25?

5) What percent of 25 is 75?

6) What percent of 60 is .45?

7) What number is 99% of 250?

8) 100 is 75% of what number?

9) 100 is 20% of what number?

10) .25 is what percent of 1.50?

11) What is $2\frac{1}{2}$% of 250?

12) What number is $17\frac{2}{3}$% of 600?

13) 21.9 is what percent of 21.9?

14) 24 is $33\frac{1}{3}$% of what number?

15) 18 is what percent of 200?

16) 100 is what percent of 75?

17) 75 is what percent of 100?

18) .03 is $\frac{3}{4}$% of what number?

19) What is $66\frac{2}{3}$% of 114?

20) What percent of 96 is 80?

21) 12 is what percent of 32?

22) What is $6\frac{1}{2}$% of $400?

23) .375 is what percent of 1.25?

24) 93 is $66\frac{2}{3}$% of what number?

25) What is $137\frac{1}{2}$% of 48?

THE "DIAGRAM" APPROACH TO PERCENTS

We have already seen that percents are numbers that *compare* certain amounts to certain base numbers.

In every percent problem there are three main parts:

AMOUNT	(A)
BASE	(B)
PERCENT	(%)

We will construct a diagram to help us do any type of percent problem.

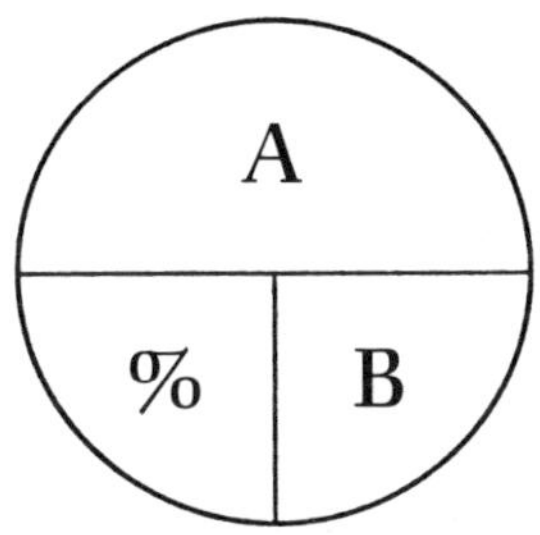

Given any two quantities to fit in the circle above, we can always find the missing value.

Example 1: Suppose both AMOUNT (A) and BASE (B) are given.

We want to find PERCENT (%).

If we hold our finger over the "%" in the circle, we have the relationship necessary to find that value.

$$\% = \frac{A}{B}$$

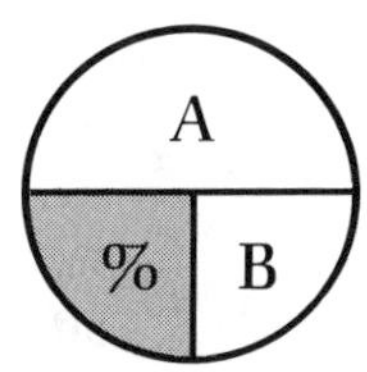

All we need do is divide A by B. (Recall: this is the ratio form for percents.)

Example 2: Suppose % and B are given. We want to find A.

If we hold our finger over the A in the circle, we have the relation necessary to find the AMOUNT (A).

$$A = \% \times B$$

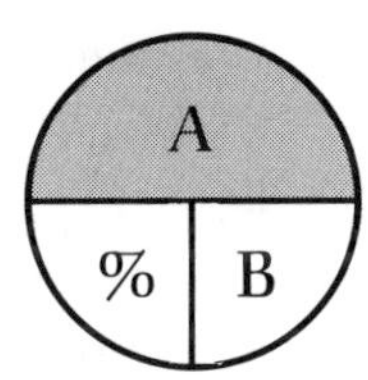

All we need do is multiply % times B (BASE).

Example 3: Suppose PERCENT (%) and AMOUNT (A) are given, but we want to find the BASE (B).

If we hold our finger over the B in the circle, we get the relation necessary to find the B value.

$$B = \frac{A}{\%}$$

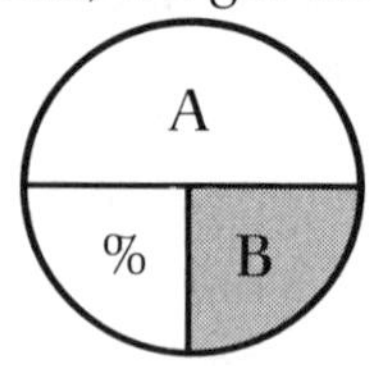

All we need do is divide the AMOUNT (A) by the PERCENT (%).

IN ALL PROBLEMS, WE MUST BE AWARE OF WHAT FORM PERCENT IS NEEDED. SOMETIMES WE WANT IT IN DECIMAL FORM, OTHER TIMES NOT.

Example 4: What is 25% of 88?

We are given PERCENT % = 25% and BASE B = 88 (since it appears after "of")

We want to find AMOUNT (A).

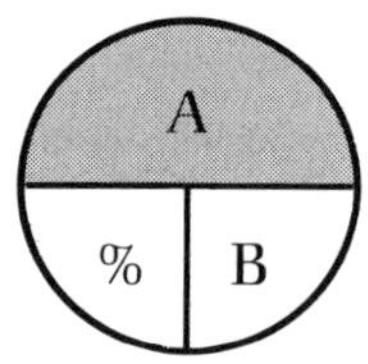

$A = \% \times B$ (from diagram)

$A = .25 \times 88 = 22.$

Example 5: What percent of 60 is 12?

BASE B = 60

AMOUNT A = 12

We need to find PERCENT (%).

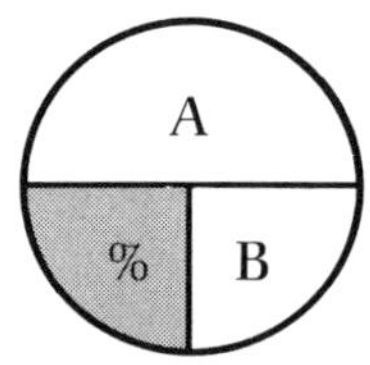

$\% = \frac{A}{B}$ (from diagram)

$\% = \frac{12}{60} = .2$ (we need to change .2 to 20%)

$\% = 20\%.$

Example 6: 25 is 20% of what number?

AMOUNT A = 25

PERCENT = 20% = .20

We need to find BASE (B)

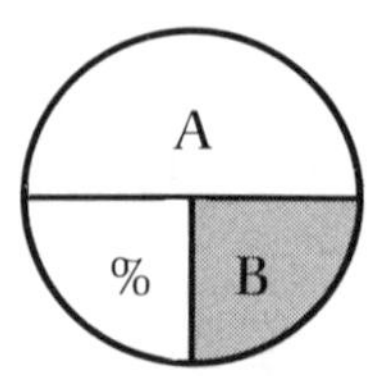

$B = \frac{A}{\%}$ (from diagram)

$B = \frac{25}{.20} = 125.$

Redo all the exercises in USE OF PERCENTS with this new "diagram" shortcut. Decide which method you like better.

APPLICATIONS USING PERCENTS

Word problems can also be solved using our ratio format, but it is sometimes tricky deciding which numbers are the BASE or AMOUNT. The usual rule is that whatever number is associated with the word *"of"* is the BASE, and whatever number is associated with the word *"is"* becomes the AMOUNT. But not every problem makes this obvious.

Here are some examples to help you. Remember: in calculations PERCENTS MUST BE CHANGED TO DECIMAL FORM. PERCENT always refers to the same thing as AMOUNT. If PERCENT refers to an increase, then AMOUNT refers to an increase. If PERCENT refers to a reduction, then AMOUNT refers to a reduction.

Example 1: Jack has read 480 pages of a 750-page assignment. What percent of the assignment has he read?

Since the question is about PERCENT, that is the unknown.

We set up the ratio. The BASE is the number after the word "of". It is obvious here. The BASE is 750. The AMOUNT is 480.

$$\frac{\text{AMOUNT}}{\text{BASE}} = \frac{480}{750} = \% \qquad \frac{480}{750} = 750\overline{)480} = .64 = 64\%$$

The answer is 64% of the assignment.

Example 2: A 130-lb. woman contains 70 pounds of water.

What percent water is the woman?

This problem has tricky wording. There is no mention of the word "of" or of the word "is" relating the numbers.

We need to find the PERCENT.

Here we use common sense to put the BASE = 130 (the larger value) and the AMOUNT = 70.

We set up the ratio:

$$\frac{\overset{\text{AMOUNT}}{70}}{\underset{\text{BASE}}{130}} = .53846 \text{ etc.} = 53.8\% \text{ (rounded off to 1 decimal place)}$$

The answer is 53.8% water.

Example 3: A lovely French wine has 3 ounces of alcohol for every 25 ounces. What percent alcohol is the wine?

Again, the wording is tricky. The word "*of*" in "3 ounces of alcohol" does NOT refer to the BASE. We again use common sense to choose the BASE as 25 and AMOUNT as 3.

The ratio is $\dfrac{\text{AMOUNT}}{\text{BASE}} = \dfrac{3}{25} = \%$

$\dfrac{3}{25} = 25\overline{)3} = .12 = 12\%$

The answer is 12%.

Example 4: Muscles in the average human being account for 40% of the total body weight. How much do the muscles of a 150-lb. man weigh (on the average)?

The BASE is 150.

The PERCENT is 40%.

We don't know AMOUNT (call it "n").

$\dfrac{\text{AMOUNT}}{\text{BASE}} = \dfrac{n}{150} = 40\% \text{ (or .4)}$

Therefore: $\dfrac{n}{150} = \dfrac{.4}{1}$

Cross-multiplying we get: $n = .4 \times 150 = 60$

The answer is 60 pounds.

(If we used the "diagram" approach then $A = \% \times B$ so $A = 40\% \times 150 = .40 \times 150 = 60$ lbs.)

Example 5: The population of a certain city was 235,780 at the last census. If the population increased by 2.5% over the year, what is the population today? (round off to nearest person)

This application is a bit tricky.

The percent given is the *increase* percent.

The base is given.

Yet we are asked to find the new population. This involves finding the *increased* AMOUNT and adding it to the original population.

The *increase* percent is 2.5%.

The BASE is 235,780 people. (The *base* is often the original number.)

We are going to find the *increase* AMOUNT. (Call it "n".)

Using the ratio form we get: $\frac{\text{AMOUNT}}{\text{BASE}} = \frac{n}{235{,}780} = 2.5\%$ (or .025)

or by "diagram" approach:

$A = \% \times B$

$n = 2.5\% \times 235{,}780 = .025 \times 235{,}780 = 5894.5$

n = 5894.5 which rounds off to 5895 people.

The *increase* amount is 5895.

To get the population today we add the increase to the last census (BASE number).

The total population is 235,780 + 5895 = 241,675 people.

Another approach to this application is to think of the new population as 100% + 2.5% or 102.5% of the last census.

Then the total new population PERCENT is 102.5%.

The BASE is still 235,780 people.

The AMOUNT will represent the new population (call it "n").

Using $A = \% \times B$ our example gives us:

$n = 102.5\% \times 235{,}780 = 1.025 \times 235{,}780 = 241{,}674.5$ or 241,675 people (rounded off)

Example 6: 67.5% of your weekly salary is $230 for rent. What is your weekly salary?

The AMOUNT is 230 (it appears after the word "is").

The PERCENT is 67.5%.

We don't know the BASE (call it "n").

Using the "diagram" approach:

$$B = \frac{A}{\%}$$

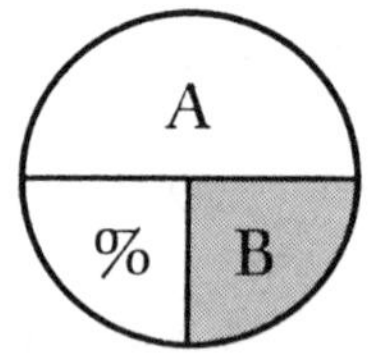

$$n = \frac{230}{67.5\%} = \frac{230}{.675} = 340.74074 \text{ etc.}$$

The answer is $340.74 (rounded off to the nearest cent).

Example 7: A student missed only 7% of the questions on an exam. If the student missed only 14 questions, how many exam questions were there?

The PERCENT is 7%.

The BASE would be the total # of questions. It is unknown. Call it "n".

The AMOUNT is 14.

Using $B = \frac{A}{\%}$ we get $n = \frac{14}{7\%} = \frac{14}{.07} = 200$.

There were 200 questions on the exam.

Example 8: A blouse is sold at a reduction of 40% off the original price. The blouse sold for $45. What was the original price?

The PERCENT is 40%, but it is the *reduction* percent.

Percent always refers to the BASE. (A percent "of" some BASE.)

The BASE is the original price. It is not known. Call it "n".

The AMOUNT is $45, but it is *not* the *reduction* amount. It is the sale price.

We must change PERCENT to sale price percent.

If 40% is the reduction, then 100% − 40% = 60% is the sale price percent.

Remember: PERCENT must always refer to the same thing as the AMOUNT. If PERCENT is increase, then AMOUNT is increase. If PERCENT refers to sales price, then AMOUNT refers to sales price. If PERCENT is decrease, then AMOUNT is decrease.

For our application, PERCENT = 60%
AMOUNT = $45
BASE = n

$$\text{BASE} = \frac{\text{AMOUNT}}{\%}$$

$$n = \frac{45}{60\%} = \frac{45}{.60} = 75$$

The original price was $75.

Example 9: A men's store buys a suit for a certain price. They sell the suit after marking up the price 48%. If the suit sells for $444, what was the cost to the store?

The AMOUNT, $444, is the sale price amount.

The sale price PERCENT would be 100% + 48% = 148%
(since selling price = cost + markup)

We don't know the BASE (cost to store). Call it "n".

$$B = \frac{A}{\%} \text{ so } n = \frac{444}{148\%} = \frac{444}{1.48} = 300.$$

The cost to the store was $300.

– WORKSHEET –

APPLICATIONS OF PERCENTS EXERCISES

Practice Set A

Round off answers to nearest cent or 1 decimal place, if needed.

1) A 140-lb. woman contains 98 pounds of water. What % of her body weight is water?

2) There are 9000 taste buds in the average human tongue. 0.7% of these are sensitive to sugar. How many taste buds are sensitive to sugar?

3) A college student working at $4.75 per hour to put herself through school is granted a 12% increase in salary. How much would she now get per hour?

4) Our graduated personal income tax system officially taxes a family with an income of $80,000 at a rate of 46%. Because of loopholes and tax writeoffs, that family only pays 26.8%. How much money is that family getting away with?

5) There are 32 billion acres of land on earth. Only 25% of this land is potentially available for crops. Of the land potentially available for crops, only 43% is presently cultivated.

   a) How much land is potentially available for crops?

   b) How much is presently being cultivated?

6) There are 225 million Americans, and about 110 million registered motor vehicles in the United States. There are only 210 million registered motor vehicles in the world.

   a) What percent of Americans have a car?

   b) What percent of all the vehicles owned in the world today are owned by Americans?

7) Of all the great earthquakes throughout the world, the following data was observed:

   Data of Earthquakes in which more than 20,000 people died:

   13 occurred in winter
   5 occurred in spring
   3 occurred in summer
   6 occurred in fall

   a) What percent occurred in winter?

   b) What percent occurred in summer?

8) Of the 271,636 immigrants admitted legally to the United States in 1975, 7308 came from Canada. What percent of the total immigrants were from Canada?

9) In 1969 the lung cancer death rate was 29.4 per 100,000 people. What percent of the population got lung cancer that year?

10) Approximately 78.08% of the air we breathe is nitrogen gas. If a container holds a liter of air closed off, how many milliliters of nitrogen gas does it contain? (liter = 1000 millileters)

11) A student missed 14 out of 150 questions on a final exam, what percentage of the problems did the student get *correct?*

12) A real estate broker receives 3% of the selling price of any house sold by any of her agents. If a woman working for her sells a home for $209,500 how much money does the broker receive for that sale?

13) Sales tax in Contra Costa County is 6½%. How much money in tax would you pay on an item that costs $299.75?

14) In a certain state the total price per gallon of gasoline is $1.28. This includes a gasoline tax of 16¢ per gallon. What percent of the total price is the gasoline tax?

15) The population of a city at the last census, 10 years ago, was 120,000. If it has increased by 24% over the last ten years, find the population today.

16) There are 23 women and 90 men on the faculty of a small college. What percent of the total faculty are the women? (round off your answer to 1 decimal place.)

17) Your weekly salary is increased by 25%. If your new salary is $450, what was your original salary?

18) On her income tax form, a businesswoman is allowed to write off her company car, valued at $12,000, as a depreciation expense. She writes off $840 the first year. What is the annual rate of depreciation she is taking as a deduction?

19) 30 milliliters of an 80% ethyl alcohol solution is mixed with 10 milliliters of pure water. What percent of the new solution is ethyl alcohol?

20) By FDA regulations, a certain cereal must contain no more than 23% sugar. An 11 ounce (311 grams) box of cereal was tested and found to contain 62 grams of sugar. Is the cereal within the FDA guidelines?

21) A family has repaid $450 of an original debt of $1800. What percent do they have yet to pay?

22) A grade school library contains 4500 books. Of these 52% are books of fiction. How many books of fiction are in the library?

23) Speedy Suzy bought a flashy new sportscar for $14,750. To get her loan at a good rate she agreed to pay 36% down, and then finance the rest. What was her downpayment?

24) As of last census figures the total world population is estimated at 4,865,000,000 people. Of these, 11.8% are children under the age of 5 years old. How many children under 5 years old are there currently in the world?

25) a) The crude birth rate in the world is given as the number of births per 1000 people. As of 1985 data it was 27.3 births per 1000 population. What is this as a percent?

b) The crude death rate in the world is 10.6 deaths per 1000 population. What is this as a percent?

c) The world growth rate is the difference between the crude birth rate and the crude death rate. What is that rate?

(Don't round off answers)

*26) In a uniformly mixed solution, 15% of the mixture was taken and found to contain 25% boric acid. What percentage of the total solution is boric acid?

27) If the probability of a certain experimental drug curing patients of a certain disease is 14%, how many patients can be expected to be cured if the drug is administered to 2800 people?

APPLICATIONS OF PERCENTS EXERCISES

Practice Set B

If necessary, round off answers to the nearest cent or to 1 decimal place.

1) There are 32 billion acres of land on earth. If the land-based Antarctic ice melted, 6.4 million acres of land would be covered with water. What percent of the earth's total land is this?

2) There are 1,037,588,000 people in the People's Republic of China by the last census. Of those, 30.7% are considered illiterate (cannot both read and write any Chinese dialect). If the government were to set up special programs for these people, how many would need to be taught?

3) If water loss amounts to 1% of body weight, a man would be thirsty. If water loss amounts to 5% of body weight, a man would collapse. If water loss amounts to 11% of body weight, a man would die. Find how much a 162 pound man could lose under each of the above circumstances.

4) When 6 pounds of sea water is evaporated, .05 pounds of salt is left. This amounts to $\frac{.05}{6} = .008$ or .8% of salt water is salt. Chemical tests on human blood yield 1.72 ounces of salt from a blood sample of 190 ounces. What percent of blood is salt? Are we saltier than sea water?

5) In 1976, there were about 750 desalting plants throughout the world. The total amount of water produced by these plants is about 750,000 gallons per day.

   a) The United States needs 300,000,000 gallons per day. What percent of the U.S. needs could be supplied by all the desalting plants of the world?

   b) The world requires about 1 trillion gallons per day. What percent of the world's needs are being supplied by the desalting plants?

6) If a discount of $18.24 is given on an item costing $76.00, what is the rate (percent) of discount?

7) You need to figure out the sales tax paid on the automobile you purchased last year so you can deduct it from your income tax. You lost the final papers on the transaction but you have the total amount. That total is $11,065.35. You know that sales tax was 6½%. How much was the car before sales tax? How much did you pay in sales tax?

8) A computer saleswoman receives commission on any computer system orders totalling more than $10,000 that she sells. Her commission is 3% for any order worth between $10,000 and $50,000, and 7% for any orders worth between $50,001 and up. Last month was very profitable. She sold three systems valued at $12,5000, $93,000 and $46,000. How much did she make in commission alone last month?

9) You have $800 in your savings account. After paying house taxes you have $75 left in the account. What percent of your savings went to pay the house taxes?

10) Seven-tenths of a percent of the entrants in a local radio show contest won prizes. If 35 people won prizes, how many entrants were there?

11) A large gift box of chocolates contains 30% nut chews, 50% dark chocolate butter creams and the rest consists of 18 fruit filled creams. How many total pieces of chocolate are in the package?

12) A real estate broker receives 3% commission on every house sold through her office. Last year alone she made $126,000 in commissions. How much money was brought into her office by the sale of houses?

13) Ammonia is usually around 15% nitrogen. If a certain cleanser is 40% ammonia, what percent of the cleanser is nitrogen?

14) The width of a river is 850 feet at the place where it is spanned by a bridge. Twenty-five percent of the bridge is on the bank of one side of the river, and fifteen percent of the bridge is on the bank of the other side of the river. How long is the bridge?

15) 100% pure gold would be marked "twenty-four carats". What percent pure is 14 carat gold? 18 carat gold?

16) How much is 10,000% of a penny?

17) A student has read 180 pages of a book for English class. He has 360 pages left to read to finish the book. What percent of the book has he read?

18) The union negotiates a two year raise package for their members. The first year's raise will be 5.5%. Then the next year the raise will be 6%. If an instructor is earning $36,000 at the time of the raise package, how much will he be making at the end of the two years?

19) The most expensive sports shoes are mink-lined golf shoes with 18-kt gold adornments and ruby tipped gold spikes. They are made by Stylo Matchmakers Ltd. of England and retail for an astronomically high amount. If they were to be reduced by 40% they would cost $5775. What is the actual cost?

20) Today there are 450 cheeses of 18 major varieties. Many are named after the small town where they are made. In France there are 240 cheeses made. What percent of all the possible kinds of cheese are made in France?

21) The United States is the country with the most number of divorces. The recent total of 1,179,000 represents 48.3% of the current annual total of marriages. What is the annual total of marriages?

22) If 40 milliliters of a 75% boric acid solution is mixed with 20 milliliters of pure water, what percent of the new solution is boric acid?

23) A college graduate has repaid $2500 of a school loan of $12,000. What percent does he have yet to pay?

24) If an oil tank contains 750 gallons, and if 240 gallons are used, what percentage of oil is left in the tank?

PERCENTS AND THE CALCULATOR

I. The important thing to remember when doing percents on the calculator is to change the percent to *decimal form*. We can do this easily by dividing by 100 on the calculator.

EVERYTIME YOU SEE A % SIGN, HIT THE (÷) BUTTON FOLLOWED BY 100. WE TYPE: (÷) 100 (=) .

Example 1: Change 24% to decimal form.

We type on the calculator: 24 (÷) 100 (=)

Example 2: Change 654% to decimal form.

We type: 654 (÷) 100 (=) on the calculator.

Example 3: Change 4.3% to decimal form.

We type: 4.3 (÷) 100 (=)

Example 4: Change $45\frac{1}{2}$% to decimal form.

We type: 1 (÷) 2 (=) (+) 45 (=) (÷) 100 (=)

II. We can also go the other way. We can change decimal numbers to *percents*.

TO CHANGE A DECIMAL TO A PERCENT WE SIMPLY MULTIPLY BY 100 AND ATTACH THE % SIGN. WE TYPE: (×) 100 (=).

Example 5: Change .0625 to percent.

We type: .0625 (×) 100 (=)

Our answer is then 62.5%

Example 6: Change 4.3 to percent.

We type: 4.3 ⊗ 100 ⊜

The answer is 430%.

Example 7: Change .0075 to percent.

We type: .0075 ⊗ 100 ⊜

The answer is .75% (or $\frac{3}{4}$%).

III. We can do any of the type of problems in our "USE OF PERCENTS" section on the calculator.

Example 8: What is 18% of 40?

This problem involves multiplying 18% times 40.

We would type on the calculator: 18 ⨸ 100 ⊜ ⊗ 40.

The answer is 7.2

Example 9: What percent of 50 is 30?

We set up our ratio of $\frac{\text{AMOUNT}}{\text{BASE}} = \frac{30}{50} = \%$

On the calculator we type: 30 ⨸ 50 ⊜ ⊗ 100 ⊜

The answer is 60%.

Example 10: 10 is 75% of what number?

Again, we set up our ratio form: we let "n" represent the BASE.

$$\frac{10}{n} = 75\%$$

By cross multiplying we get 75%n = 10

Then $n = \frac{10}{75\%}$

On the calculator we will type: 75 (÷) 100 (=)   (75% = .75)

Then the problem is $n = \frac{10}{.75}$

Then we type on the calculator: 10 (÷) .75 (=)

The answer is $13.\overline{3}$

Example 11: What is $12\frac{1}{4}\%$ of 50?

Since we can multiply without forming ratio form, we will do this directly on the calculator.

We type: 1 (÷) 4 (+) 12 (=) (÷) 100 (=) (×) 50 (=)

The answer is 6.125

Example 12: 24 is 40% of what number?

Again, we set up the ratio form: we let "n" be the BASE.

$$\frac{24}{n} = 40\%$$

Then we get $40\%n = 24$

Then $n = \frac{24}{40\%}$

On the calculator we will type 40 (÷) 100 (=)   (40% = .4)

Then the problem is $n = \frac{24}{.4}$

We then type: 24 (÷) .4 (=)

The answer is 60.

Example 13: What percent of $37\frac{1}{2}$ is 75?

Since $37\frac{1}{2}$ follows the word "of", it is the BASE. The AMOUNT is 75.

We set up our ratio of $\frac{\text{AMOUNT}}{\text{BASE}} = \frac{75}{37\frac{1}{2}}$

On the calculator we type: 75 (÷) 37.5 (=) (×) 100 (=)

The answer is 200%.

PERCENTS AND THE CALCULATOR EXERCISES

Practice Set A

Change each of the following percents to decimals on the calculator:

1) 47% 2) 36% 3) 4% 4) 26% 5) 52%

6) 125% 7) 235% 8) $12\frac{1}{2}\%$ 9) $33\frac{1}{3}\%$ 10) $66\frac{2}{3}\%$

11) 60% 12) $\frac{1}{4}\%$ 13) $1\frac{1}{2}\%$ 14) 2.75% 15) 93.5%

16) 87.25% 17) .5% 18) .0025% 19) 1.08% 20) $32\frac{7}{8}\%$

Change each of the following decimals to percents on the calculator:

21) .38 22) .54 23) .78 24) .8 25) 2.1

26) 1.45 27) .0025 28) .0375 29) .3 30) .15

31) .625 32) 3 33) $.04\frac{1}{2}$ 34) .002 35) .0075

Work the following percents exercises on the calculator:

36) What is 25% of 96?

37) What percent of 55 is 44?

38) 20 is $33\frac{1}{3}\%$ of what number?

39) What is $37\frac{1}{2}\%$ of 56?

40) What percent of 42 is 28?

41) 36 is $66\frac{2}{3}\%$ of what number?

42) What percent of 14 is 35?

43) What is $\frac{3}{4}\%$ of 80?

44) 45 is 225% of what number?

PERCENTS AND THE CALCULATOR EXERCISES

Practice Set B

Change each of the following percents to decimals on the calculator:

1) 37% 2) 48% 3) 2% 4) 23% 5) 53%

6) 175% 7) 241% 8) $37\frac{1}{2}\%$ 9) $83\frac{1}{3}\%$ 10) $1\frac{3}{4}\%$

11) 30% 12) $\frac{2}{3}\%$ 13) $6\frac{1}{2}\%$ 14) 8.25% 15) 99.25%

16) 109.7% 17) .6% 18) .0075% 19) 2.3% 20) $210\frac{3}{8}\%$

Change each of the following decimals to percents on the calculator:

21) .31 22) .64 23) .72 24) .7 25) 1.8

26) 2.33 27) .005 28) .0625 29) .1 30) .28

31) .375 32) 2 33) $.16\frac{3}{4}$ 34) .008 35) .0125

Work the following percent exercises on the calculator:

36) What is $33\frac{1}{3}\%$ of 72?

37) What percent of 75 is 15?

38) 27 is $66\frac{2}{3}\%$ of what number?

39) What is 80% of 45?

40) What percent of 65 is 13?

41) 18 is $1\frac{1}{2}\%$ of what number?

42) What percent of 10 is 24?

43) What is .6% of 150?

44) 28 is 175% of what number?

PERCENTS - REVIEW EXERCISES

Practice Set A

Change the following percents to decimal form:

1) 37% 2) 17% 3) 59% 4) 6% 5) 35%

6) $7\frac{1}{4}\%$ 7) $66\frac{2}{3}\%$ 8) 205% 9) 101% 10) .375%

Change the following decimals to percents:

11) .79 12) .03 13) .16 14) .40 15) 1.42

16) $.\overline{666}$ 17) .0375 18) .24375 19) .005 20) 4

21) Fill in the following table of values with equivalent values:

Decimal Form	Percent Form	Fraction Form
		$\frac{1}{3}$
	75%	
.0625		
		$\frac{2}{5}$
	$133\frac{1}{3}\%$	
.125		
		$\frac{3}{8}$
	48%	
.05		
		$\frac{1}{6}$
	$2\frac{3}{4}\%$	
2.3		

22) What is 45% of 270?

23) What percent of 15 is 2.5?

24) What number is $66\frac{2}{3}$% of 243?

25) 15 is $33\frac{1}{3}$% of what number?

26) 1 is 2% of what number?

27) What percent of 30 is 100?

28) What is 125% of 36?

29) 27 is what percent of 21.6?

30) 3.5 is $6\frac{1}{4}$% of what number?

31) Your weekly salary is increased by 30%. If your new salary is $520, what was your original salary?

32) At a recent rock concert where 9000 people attended the show, 6750 were under the age of 21. What percent of the people were of an age of 21 or more?

33) If $\frac{1}{2}$% of all the entrants in a local contest won prizes, and if 90 people won prizes, how many entrants were there?

34) What was the regular price of a suit if it sold for $246 after an 18% reduction in price?

35) Out of a class of 150 students, 90 students received lower than a B grade. What percent of the class got either an A or B grade?

PERCENTS - REVIEW EXERCISES

Practice Set B

Change the following percents to decimal form:

1) 39% 2) 19% 3) 41% 4) 2% 5) 76%

6) $8\frac{3}{4}\%$ 7) $33\frac{1}{3}\%$ 8) 198% 9) 202% 10) .625%

Change the following decimals to percents:

11) .69 12) .01 13) .23 14) .6 15) 2.37

16) $1.\overline{666}$ 17) .0875 18) .4525 19) .0025 20) 6

21) Fill in the following table of values with equivalent values:

Decimal Form	Percent Form	Fraction Form
		$\frac{2}{3}$
	50%	
.0875		
		$\frac{3}{5}$
	$266\frac{2}{3}\%$	
.625		
		$\frac{5}{16}$
	72%	
.04		
		$\frac{5}{6}$
	$1\frac{1}{2}\%$	

22) What is 35% of 70?

23) What percent of 36 is 12?

24) What number is $33\frac{1}{3}\%$ of 48?

25) 18 is $66\frac{2}{3}\%$ of what number?

26) 2 is 1% of what number?

27) What percent of 40 is 100?

28) What is $212\frac{1}{2}\%$ of 40?

29) 75 is what percent of 250?

30) 3.3 is $5\frac{1}{2}\%$ of what number?

31) If the University accepts only $2\frac{1}{2}\%$ of all the people who apply for admission to the School of Business and they accepted 250 students this year, how many applied for admission to the School of Business?

32) If the population of a certain town was 12,480 at the last census, and it increased by 2.5% in one year, what was the population after that year?

33) A coat is marked up by 60% and sold for $480. What was the original cost of the coat?

34) What percent pure gold is 21 carat gold? (Recall: 24 carats = 100% pure gold)

35) If 25 ounces of a 40% hydrochloric acid solution is mixed with 5 ounces of water, what percent of the new solution is hydrochloric acid?

PERCENTS: REVIEW OF IMPORTANT WORDS

Here is a checklist of words that are used in dealing with percents. See if you can identify each one and define it to yourself.

Percent

Base

Amount

Ratio Form

Decimal Form

Fraction Form

SAMPLE PERCENTS EXAM #1

SHOW ALL WORK FOR FULL CREDIT.

I. Change these percents to decimals:

1) 57%

2) 352%

3) $\frac{3}{4}\%$

4) .6%

II. Change these decimals to percents:

1) .65

2) .0435

3) 4.36

4) .29875

III. Use of percents:

1) Find 30% of 50

2) What is $8\frac{1}{2}\%$ of 150?

3) What percent of 160 is 48?

4) $12\frac{1}{2}$ is what percent of 200?

5) 17 is 25% of what number?

6) 26 is $6\frac{1}{2}\%$ of what number?

IV. Applications:

1) Sales tax in Contra Costa County is $6\frac{1}{2}\%$. How much sales tax would you pay on an item that sells for $249.49?

2) Your weekly salary of $900 is increased by 20%. What will your new salary be?

3) A student missed 12 out of 80 questions on an exam. What percent of the problems did the student get correct?

4) A men's sportscoat sold at a reduction of $33\frac{1}{3}\%$ off the original price. The sportscoat sold for $120. What was the original price?

5) If 20 milliliters of a 50% ammonia solution is mixed with 30 milliliters of pure water, what percent of the new solution is ammonia?

SAMPLE PERCENTS EXAM #2

SHOW ALL WORK FOR FULL CREDIT.

I. Change these percents to decimals:

1) 24%

2) 415%

3) $1\frac{1}{2}\%$

4) .375%

II. Change these decimals to percents:

1) .57

2) .08375

3) 2.56

4) .341

III. Use of percents:

1) Find $33\frac{1}{3}\%$ of 48

2) What is $2\frac{1}{4}\%$ of 20?

3) What percent of 216 is 36?

4) $7\frac{1}{2}$ is what percent of 150?

5) 22 is 25% of what number?

6) 30 is $7\frac{1}{2}\%$ of what number?

IV. Applications:

1) Sales tax in New York City is 8%. How much sales tax would you pay on an item that sells for $1009.39?

2) If the new Federal income tax laws change, a family with an income of $80,000 can go from paying 46% in taxes to 26%. How much more money would the family have under the proposed new tax law?

3) A family has repaid $1200 of an original $5000 debt. What percent do they have yet to pay?

4) A fine china shop buys a certain brand of china for a wholesale price. They sell the china after marking up the price 75%. If a set of the china sells for $1200, what was the cost to the store? (Round off to the nearest cent.)

5) The population of a certain city was 352,400 at the last census. If the population increased by $1\frac{3}{4}\%$ over the year, what is the population after the year? (Round off to nearest person.)

"30 LITERS...30 LITERS...!, NOT 30 GALLONS"

CHAPTER 5:
THE METRIC SYSTEM

Contrary to our own clumsy inch-quart-pound system of measurement, the metric system is a more elegant system where the measurements are all related to one another. The prefix names tell us these relationships. It is a convenient system based on our base-10 system of numeration. Changing from one metric measurement to another requires no more than moving decimal points.

BASICS OF THE METRIC SYSTEM: LENGTH

The standard unit of length is the *meter* (abbreviated as m.). A meter is a little longer than a yard (approximately 39.37 inches). The meter is the basis of all units of length in the metric system.

If we take the meter and divide it into 10 equal parts, we have what is called a "*decimeter*". "Deci" literally means "one-tenth". Here is a line of that actual length:

1 dm = |————————————————————————|

If we divide the decimeter into 10 equal subunits, we have the next smaller unit of length called the "*centimeter*". Basically we have taken our original measure, the meter, and divided it twice into 10 equal intervals, or 10 × 10 smaller pieces. "Centi" literally means "one-hundredth".

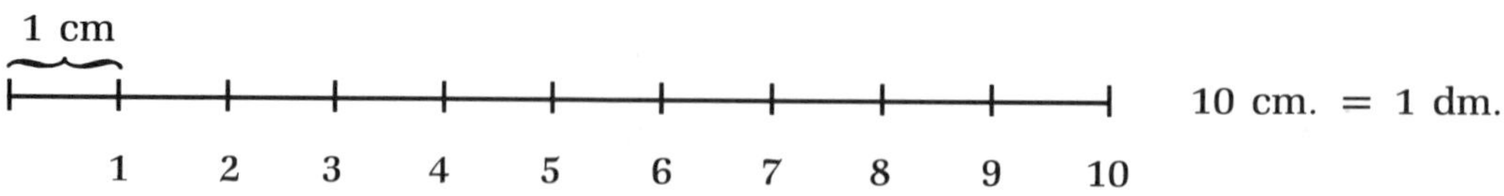

Now if we take the centimeter and divide it into 10 equal pieces, we have what is called a "*millimeter*". The original meter has been divided three separate times into 10 equal pieces, or 10 × 10 × 10 (or 1000) smaller units. "Milli" literally means "one-thousandth".

We can continue in this manner getting smaller and smaller units of measure, all based on the meter.

Just as we used the meter as the larger measure that we subdivided, so can we use it as the smaller unit that we piece together to get larger ones. Ten meters pieced together would give us what is called a "*dekameter*". "Deka" literally means "ten times."

One hundred meters pieced together forms what is called a "*hectometer*". "Hecto" literally means "one hundred times".

One thousand meters pieced together gives us the familiar "*kilometer*". "Kilo" literally means "one thousand times".

The following chart will summarize the most commonly used measures of length.

Measure	**Symbol**	**Prefix Meaning**	**Relation to 1 Meter**
*milli*meter	mm.	$\frac{1}{1000} = .001$	1000 mm. = 1 m.
*centi*meter	cm.	$\frac{1}{100} = .01$	100 cm. = 1 m.
*deci*meter	dm.	$\frac{1}{10} = .1$	10 dm. = 1 m.
meter	m.	(1)	1 m. = 1 m.
*deka*meter	dam.	10	1 dam. = 10 m.
*hecto*meter	hm.	100	1 hm. = 100 m.
*kilo*meter	km.	1000	1 km. = 1000 m.

To convert from one measure on our chart to another is simply a matter of moving decimal points. We will see these shortcuts after looking at some problems worked out the long way, with conversion factors.

Example 1: Change 2.4 m. to cm.

Since 1 m. = 100 cm., we can create two fractional equivalents of 1.

$$\frac{100 \text{ cm.}}{1 \text{ m.}} = 1 \quad \text{or} \quad \frac{1 \text{ m.}}{100 \text{ cm.}} = 1$$

Since we want to change from meters to centimeters, we use the fraction with cm. on top. We can then multiply as follows:

$$2.4 \text{ m} = \frac{2.4 \text{ m.}}{1} \times \underbrace{\frac{100 \text{ cm.}}{1 \text{ m.}}}_{1} = \frac{2.4 \not{\text{m}}.}{1} \times \frac{100 \text{ cm.}}{1 \not{\text{m}}.} =$$

$$2.4 \times 100 \text{ cm.} = 240 \text{ cm.}$$

NOTE: Meters cancelled leaving 2.4 × 100 cm.

Our result is that 2.4 m = 240 cm.

We can see a shortcut if we realize that the answer could be gotten by moving the decimal point *two places to the RIGHT*.

In the chart, centimeter is two steps UP from meter.

$$2.4 \text{ m} = 2.\underset{\smile\smile}{40}\text{ cm.} = 240 \text{ cm.}$$

Example 2: Change 8700 mm. to m.

Since 1 m. = 1000 mm., we can create the fractional equivalent to 1:

$$\frac{1 \text{ m.}}{1000 \text{ mm.}} = 1$$

NOTE: The m. is on top since we are converting to it.

We multiply:

$$8700 \text{ mm.} = \frac{8700 \text{ mm.}}{1} \times \frac{1 \text{ m.}}{1000 \text{ mm.}} = \frac{8700}{1000} \times 1 \text{ m.} = 8.7 \text{ m.}$$

8700 mm. = 8.7 m.

We can see a shortcut if we realize that the answer could be gotten by moving the decimal point *three places to the left.* In the chart, millimeter is *three* steps DOWN from meter.

8700 mm. = 8 7 0 0 m = 8.700 m = 8.7 m.

> SHORTCUTS: In converting from one metric measure of length to another, determine in what direction in the chart you are moving. Identify the metric measure you are starting with in the chart. If you are moving UP to another measure, you move the decimal point to the RIGHT the same number of places as the steps UP the chart. If you are moving DOWN the chart to another measure, you move the decimal point to the LEFT the same number of places as the steps DOWN the chart.

Example 3: Change 9.25 dm. to mm.

From dm. we are *two* steps UP the chart to mm. Therefore, we move the decimal point *two* places to the RIGHT.

9.25 dm = 9. 2 5 mm = 925 mm.

Example 4: Change 3 km. to m.

From km. to m. we are *three* steps UP the chart. We move the decimal point *three* places to the RIGHT.

(NOTE: 3 = 3.000)

3 km. = 3. 0 0 0 m. = 3000 m.

3 km. = 3000 m.

Example 5: Change 75000 dm. to km.

From dm. we are *four* steps DOWN the chart to km.

We move the decimal point *four* places to the LEFT.

75000 dm. = 7 5 0 0 0. km = 7.5000 km. = 7.5 km.

BASICS OF THE METRIC SYSTEM: LENGTH EXERCISES

Practice Set A

Work the following conversions:

1) Change 1800 cm. to m.

2) Change 45,000 m. to km.

3) Change 78.6 dm to cm.

4) Change 17 hm. to m.

5) Change 80 km. to m.

6) Change 56 dam. to m.

7) Change 432.1 dam. to mm.

8) Change 10 km. to mm.

9) Change 482,803,000 cm. to km.

10) Change 185m. to dam.

11) Change 5.3 mm to cm.

12) Change 2.8 dam. to km.

13) Change 2840 hm. to km.

14) Change 2.4 m to cm.

15) Change .028 hm to mm.

16) Change 8.25 dm. to cm.

17) Change 2500 cm. to m.

18) Change 1500 mm. to dm.

19) Change 1,463,000 mm. to dam.

20) Change .00586 km. to mm.

BASICS OF THE METRIC SYSTEM: LENGTH EXERCISES

Practice Set B

Work the following conversions:

1) Change 2700 cm. to m.

2) Change 58,700 m. to km.

3) Change 98.2 dm. to cm.

4) Change 19 hm. to m.

5) Change 83 km. to m.

6) Change 147 dam. to m.

7) Change 953.1 dam. to mm.

8) Change 30 km. to mm.

9) Change 2,876,000,000 cm. to km.

10) Change 375 m. to dam.

11) Change 2.8 mm. to cm.

12) Change 9.86 dam. to km.

13) Change 4710 hm. to km.

14) Change 6.3 m. to cm.

15) Change .046 hm. to mm.

16) Change 11.03 dm. to cm.

17) Change 4800 cm. to m.

18) Change 14,000 mm. to dm.

19) Change 23,486,000 mm. to dam.

20) Change .00138 km. to mm.

BASICS OF THE METRIC SYSTEM: WEIGHT

The standard metric measure of weight is the *gram*, abbreviated as gm. To get an idea of how heavy this is, imagine a cube, 1 cm. on each side. Fill that cube with water. The weight of that amount of water would be 1 gram. Here is a scale drawing of the cube:

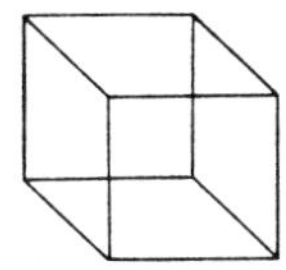

1 teaspoonful of water weighs about 5 grams.

1 nickel weighs about 5 grams.

The two prefixes that are most commonly used with *gram* are "milli" and "kilo".

Here is a chart summarizing the weight measures:

Measure	Symbol	Prefix Meaning	Relation to 1 Gram
*milli*gram	mg.	$\frac{1}{1000} = .001$	1000 mg. = 1 g.
*centi*gram	cg.	$\frac{1}{100} = .01$	100 cg. = 1 g.
*deci*gram	dg.	$\frac{1}{10} = .1$	10 dg. = 1 g.
gram	g.	(1)	1 g. = 1 g.
*deka*gram	dag.	10	1 dag. = 10 g.
*hecto*gram	hg.	100	1 hg. = 100 g.
*kilo*gram	kg.	1000	1 kg. = 1000 g.

Changing between these weight measures is as simple as how we converted our length measures. We will use the same shortcut for the weight conversion problems.

Example 1: Change 4000 g. to kg.

To go from gram to kilogram we move DOWN the chart three steps, so we move the decimal point *three* places to the LEFT.

4000 g. = 4 0 0 0. kg. = 4 kg.

Example 2: Change 2.5 kg. to mg.

To go from kg. to mg. we move UP the chart *six* places.

Therefore, we move the decimal point *six* places to the RIGHT.

2.5 kg. = 2. 5 0 0 0 0 0 mg. = 2,500,000 mg.

Example 3: Change 11.8 dag. to dg.

From dag. it is *two* steps UP the chart.

We move the decimal point *two* places to the RIGHT.

11.8 dag. = 11. 8 0 dg. = 1180 dg.

Example 4: Change 5.25 mg. to g.

From mg. it is *three* steps DOWN the chart.

We move the decimal point three places to the LEFT.

5.25 mg. = 0 0 5.25 g. = .00525 g.

BASICS OF THE METRIC SYSTEM: WEIGHT EXERCISES

Practice Set A

Work the following conversions:

1) Change 4.8 g. to mg.

2) Change 850,000 g. to kg.

3) Change 19,810 mg. to g.

4) Change 3 kg. to mg.

5) Change 145,000 mg. to kg.

6) Change 37.9 g. to kg.

7) Change 1800 dag. to hg.

8) Change 1800 dag. to cg.

9) Change 1800 dag. to kg.

10) Change 2.5 kg. to g.

11) Change 3700 mg. to g.

12) Change $\frac{1}{2}$ kg. to g.

13) Change 19,500 mg. to kg.

14) Change 450 g. to kg.

15) Change 17,700 cg. to dag.

16) Change 1 g. to kg.

17) Change 1500 g. to hg.

18) Change 475 cg. to mg.

19) Change 19.7 dg. to mg.

20) Change 5,463,000 mg. to kg.

BASICS OF THE METRIC SYSTEM: WEIGHT EXERCISES

Practice Set B

Work the following. conversions:

1) Change 5.6 g. to mg.

2) Change 975,000 g. to kg.

3) Change 24,600 mg. to g.

4) Change 4 kg. to mg.

5) Change 793,000 mg. to kg.

6) Change 22.3 g. to kg.

7) Change 6300 dag. to hg.

8) Change 6300 dag. to cg.

9) Change 6300 dag. to kg.

10) Change 3.25 kg. to g.

11) Change 7600 mg. to g.

12) Change $\frac{1}{4}$ kg. to g.

13) Change 32,900 mg. to kg.

14) Change 975 g. to kg.

15) Change 18,500 cg. to dag.

16) Change 4 g. to kg.

17) Change 1800 g. to hg.

18) Change 625 cg. to mg.

19) Change 32.4 dg. to mg.

20) Change 7,428,000 mg. to kg.

BASICS OF THE METRIC SYSTEM: VOLUME AND CAPACITY

There is a need to distinguish between *the amount of space that an object occupies* versus *the amount of substance that a container holds*. The amount of space that an object occupies is referred to as a VOLUME measurement. When we think of volume many of us will remember the simple formula for the volume of a box (length times width times height). This measures the space taken up by the box. It usually has a measure such as "cubic inches" or "cubic feet". On the other hand, the amount of substance that a container will hold is considered a CAPACITY measurement. If we look at a typical can of food in the grocery store in this country and read the label, it will be measured in "fluid ounces." "Fluid ounces" measures CAPACITY, whereas "cubic inches" or "cubic feet" measures VOLUME.

The basic metric measure of CAPACITY is the "liter". The liter measures the amount of substance that a container of a certain size will hold. Like all other metric measures, it can be divided into smaller parts (deciliter, centiliter and milliliter), or pieced together to form larger quantities (like dekaliter, hectoliter or kiloliter).

"Liter" and "milliliter" are the ones most commonly used. Here is the chart showing all the gradations:

Measure	Symbol	Prefix Meaning	Relation to 1 Gram
*milli*liter	ml.	$\frac{1}{1000} = .001$	1000 ml. = 1 l.
*centi*liter	cl.	$\frac{1}{100} = .01$	100 cl. = 1 l.
*deci*liter	dl.	$\frac{1}{10} = .1$	10 dl. = 1 l.
liter	l.	(1)	1 l. = 1 l.
*deka*liter	dal.	10	1 dal. = 10 l.
*hecto*liter	hl.	100	1 hl. = 100 l.
*kilo*liter	kl.	1000	1 kl. = 1000 l.

Converting from one capacity measure to another is as simple as converting between length measures or weight measures. We use the chart and count the number of steps UP or DOWN the chart and move the decimal point RIGHT or LEFT.

Example 1: Change 7.8 liters to milliliters.

To go from liters to milliliters, we move *three* steps UP the chart, therefore we move the decimal point three places to the RIGHT.

7.8 l. = 7. 8 0 0 ml. = 7800 ml.

Example 2: Change .054 kl. to l.

To go from kl. to l. we move UP the chart three places. We need to move the decimal point three places to the RIGHT.

.054 kl. = . 0 5 4 liters = 54 l.

Example 3: Change 1850 cl. to dal.

We move down the chart three places. We move the decimal point three places LEFT.

1850 cl. = 1 8 5 0 . = 1.85 dal.

VOLUME

If we think of the formula for the volume of a box which is (length) × (width) × (height), we see that we always have three dimensions. Our answer will involve three length measures that are multiplied.

In our English system of measure we have measures like: feet × feet × feet or inch × inch × inch. Written other ways: ft. × ft. × ft. = $(\text{ft.})^3$ = cubic feet or in. × in. × in. = $(\text{in.})^3$ = cubic in.

In our metric system we have measures like:

cm. × cm. × cm. = $(\text{cm.})^3$ = cubic centimeters (abbreviated cm^3.)

or m. × m. × m. = $(\text{m.})^3$ = cubic meters (abbreviated m^3.)

Example 1: Find the volume of a box whose length is 8 cm., width is 3 cm. and height is 10 cm.

The volume = l × w × h

V = 8 cm. × 3 cm. × 10 cm. = (8 × 3 × 10)cm^3. = 240 cm^3.

Example 2: Change 1 m^3. to cubic centimeters.

Since 1 m. = 100 cm. then 1 m^3. = 1 m. × 1 m. × 1 m.

= (100 cm.) × (100 cm.) × (100 cm.) = 1,000,000 cm^3.

We substituted 100 cm. for 1 m. three times.

1 m^3. = 1,000,000 cm^3.

Example 3: Change 4 dm^3. to cubic centimeters.

Since 1 dm. = 10 cm. then 4 dm^3. = 4 × (dm. × dm. × dm.)

= 4 × (10 cm. × 10 cm. × 10 cm.)

= 4 × 1000cm^3. = 4000 cm^3.

4 dm^3. = 4000 cm^3.

NOTATION: 5 cm^3. means 5 × (cm × cm × cm)

It does NOT mean 5 cm. × 5 cm. × 5 cm., which would be written as (5 cm.)^3.

Example 4: Change $(5 \text{ cm.})^3$ to cubic meters.

Since 1 cm. = .01 m and $(5 \text{ cm.})^3 = 5 \text{ cm.} \times 5 \text{ cm.} \times 5 \text{ cm.}$

then $(5 \text{ cm.})^3 = (5 \times .01 \text{ m.}) \times (5 \times .01 \text{ m.}) \times (5 \times .01 \text{ m.})$

$= 125 \times .000001 \text{ m}^3.$

$= .000125 \text{ m}^3.$

BASICS OF THE METRIC SYSTEM: VOLUME AND CAPACITY EXERCISES

Practice Set A

Work the following conversions:

1) Change 4.5 l. to ml.

2) Change 18,000 l. to kl.

3) Change .125 l. to cl.

4) Change .089 kl. to l.

5) Change .00875 hl. to ml.

6) Change $\frac{1}{2}$ l. to ml.

7) Change 15,000 ml. to l.

8) Change 4850 dl. to hl.

9) Change 2 dal. to cl.

10) Change .125 l. to ml.

11) Change 1 cm^3. to m^3. (cubic meters)

12) Change 5 m^3. to dm^3. (cubic decimeters)

13) Change 1000 km^3. to m^3. (cubic meters)

14) Change $\frac{1}{2}$ m^3. to cm^3. (cubic centimeters)

15) Change .25 cm^3. to mm^3. (cubic millimeters)

BASICS OF THE METRIC SYSTEM: VOLUME AND CAPACITY EXERCISES

Practice Set B

Work the following conversions:

1) Change 5.6 l. to ml.

2) Change 25,000 l. to kl.

3) Change .125 l. to cl.

4) Change .048 kl. to l.

5) Change .00375 kl. to ml.

6) Change $\frac{3}{4}$ l. to ml.

7) Change 37,000 ml. to l.

8) Change 5370 dl. to hl.

9) Change 3.25 dal. to cl.

10) Change .025 l. to ml.

11) Change 2 cm^3. to mm^3.

12) Change 8 m^3. to cm^3.

13) Change 35,000 km^3. to m^3.

14) Change $\frac{3}{8}$ m^3. to cm^3.

15) Change .125 cm^3. to mm^3.

ADDITIONAL METRIC SYSTEM EXERCISES

How much intuition do you have about the metric system? In each of the following circle the most appropriate answer.

1) The average man has a height of: a) 4 meters b) 4 cm. c) 1.75 m.

2) A feather weighs: a) 1 kg. b) 1 g. c) 100 g.

3) In an hour a person can walk: a) 1 meter b) 1 km. c) 1 cm.

4) A pumpkin weighs: a) 1 kg. b) 100 kg. c) 3.2 kg.

5) A person on a diet can eat the following weight of beef daily:
a) 1 g. b) 1 kg. c) 100 g.

6) A five year old child of normal height would be how tall?
a) 10 cm. b) 10 m. c) 100 cm.

7) You live over a mile from campus. You live how far from school?
a) 1 km. b) 900 m. c) 2 km.

8) A fifth of wine measures roughly: a) .75 liters b) 5 liters c) 500 ml.

9) What distance does a car travel in an hour? (assume highway speed)
a) 90 meters/hr. b) 90 km./hr. c) 900 meters/hr.

10) This sheet of paper has length: a) .5 meters b) 100 mm. c) 28 cm.

11) A nickel weights about: a) 5 mg. b) 5 kg. c) 5 g.

12) A female model may weigh: a) 275 kg. b) 53 kg. c) 1250 grams

13) A small bag of sugar from the grocery store would weigh:
a) 2 kg. b) 250 g. c) 20 kg.

14) A football field is as long as: a) 91 meters b) 1 km. c) 11,000 mm.

15) A runner in a marathon race went: a) 42 km. b) 13.1 km. c) 2650 m.

Fill in the missing blanks with the correct number. Then check to see that you are correct, and MEMORIZE these approximations.

16) A kilometer is a little longer than ____________.

17) A meter is about ____________% longer than a yard.

18) A centimeter is a little shorter than ____________ an inch.

19) A liter holds a little more than a ____________.

20) A kilogram is about ____________% heavier than 2 pounds.

21) A milliliter is about ____________ of a teaspoon.

22) Complete the following equivalences within the Metric System:

3 m. is ______ mm.	150 cm. is ______ m.
3 km. is ______ m.	1500 mm. is ______ cm.
3 m. is ______ km.	1500 mm. is ______ m.
91.44 m. = ______ cm.	3500 mm. = ______ km.

23) Change $\frac{2}{5}$ kilometers to meters.

24) Change 2.25 meters to centimeters.

25) Find the number of millimeters in $3\frac{1}{2}$ cm.

26) How many decimeters is the same as .0007 km.?

27) 450 grams is ____________ kilograms.

28) A swimming pool has dimensions 30 meters × 10 meters × 2.5 meters. What is the volume of the pool in cubic meters?

29) A whale weighs about 10,000,000 grams. How many kg. is this?

30) A fly weighs .0007 kg. How many grams is this?

Metric Experiments:

31) Measure the length of your room at home with a *metric* ruler. If the dimensions are not exactly whole numbers, mark the spot in the room where the largest whole number in that measurement falls.

Try to set your pace at exactly 1 meter. You will have to practice this several times to get a feeling for this measurement. We will call this your "meter pace."

a) Measure the length and width of two other rooms with your "meter pace".

b) Now measure your rooms with the metric ruler. Did your pace measurement agree?

c) Convert your answer into "yards" by adding 10% of your meter answer to the meter answer.

32) We eat food to enable us to do work in our lives. The food we eat is measured by the "potential energy" for work that it provides. Work is defined by scientists as "force × distance".

If you lift 100 grams a distance of 1 meter, that activity requires the energy of about "1 joule".

A nutritionist's calories is really a "kilocalorie" and measures 4 joules.

The following table gives various activities with their caloric equivalents.

ACTIVITY	# MINUTES PERFORMED	# KILOCALORIES USED	# JOULES USED
Jogging	60	600	
Biking (7 mph)	60	390	
Swimming (30 yds/min)	60	510	
Eating	60	100	
Walking (briskly)	60	312	
Standing	60	120	
Stepping	60	450	

a) Fill in the "joules" column.

b) Write down as many activities from the table that you perform in a day. Also write down the time duration for each activity.

c) Use the table to calculate the number of kilocalories you would use for each activity.

d) Add a "basal metabolism" calories amount of ____________ to your answer. (To calculate your "basal metabolism" amount, multiply your weight in pounds by 11. If you are a woman, deduct 10% from that amount.)

Now you know how many kilocalories you are allowed to consume in a day.

33) You want to lay tile in your bathroom. The room measures 250 cm. by 300 cm. How many square tiles 20 cm. on each side are needed?

34) A tablespoon holds approximately 15 ml. of liquid. What are the least number of tablespoonfuls to fill a liter bottle?

35) An "are" is the basic measure of land area. An "are" is 100 square meters (100 m.^{2} = 100 m. × m. or 10 m. × 10 m.)

a) Using the meaning of the prefix "hecto", define the "hectare".

b) How many hectares in a square kilometer?
(HINT: 1 km^2. = 1 km. × 1 km. = 1000 m. × 1000 m.)

TABLE OF ENGLISH MEASURES

Here is the table of equivalent English measurements for LENGTH, WEIGHT and VOLUME/CAPACITY. Most of these you will be familiar with, but there are some measures that are not used very often.

Length

1 foot = 12 inches

1 yard = 3 feet

1 yard = 36 inches

1 rod = 16.5 feet = 5.5 yards

1 mile = 5280 feet = 1760 yards

1 nautical mile = 6080 feet

1 fathom = 6 feet

Weight

1 pound = 16 ounces

1 ounce = 437.5 grains

1 ton = 2000 pounds

Volume/Capacity

1 cubic foot = 1728 cubic inches

1 cubic yard = 27 cubic feet

1 gallon = 231 cubic inches

1 quart = 57.75 cubic inches

1 pint = 28.875 cubic inches

1 gallon = 128 ounces (fluid ounces)

1 quart = 32 ounces (fluid ounces)

1 pint = 16 ounces (fluid ounces)

1 cut = 8 ounces (fluid ounces)

1 tbsp. = 3 tsp. = $\frac{1}{2}$ fluid ounce

CONVERSIONS BETWEEN THE METRIC AND ENGLISH SYSTEMS OF MEASUREMENT

The following charts give the values to be used when changing from metric to English and vice versa. All these values are *approximate* since the measures of the two systems are not exactly interchangeable. All numbers have been rounded off to help simplify the process.

Metric To English

1 meter = 39.37 inches

1 meter = 3.28 feet

1 meter = 1.09 yards

1 kilometer = .62 miles

1 centimeter = .39 inches

1 liter = 1.06 quarts

1 kilogram = 2.2 pounds

1 metric ton = 2204.6 pounds

English To Metric

1 inch = 2.54 centimeters

1 inch = .0254 meters

1 foot = .3 meters

1 yard = .91 meters

1 mile = 1.61 kilometers

1 gallon = 3.8 liters

1 tbsp. = 15 ml.

1 tsp. = 5 ml.

1 fluid ounce = 30 ml.

1 pound = 454 grams = .454 kg.

1 ounce = 28.35 grams

There are several ways to change from one system of measurement to another. We will do several examples using 2 of these methods (I and II). Method I replaces one measure with its equivalent value from the chart above. Method II uses "dimensional algebra". This method creates fraction equivalents of 1 and uses them in appropriate sequence to arrive at the desired conversion.

Example 1: Change 4 feet to meters.

We are changing from English to Metric, so we look at the right hand chart.

We see that 1 foot is .3 meters.

Method I: Everywhere there is the word feet, we replace it with .3 meters.

4 feet = 4 × (.3 meters) = 1.2 meters

Method II: We can also use dimensional algebra. This involves setting up conversion factors that are equivalent to 1.

The conversion factors here are: $\frac{1 \text{ foot}}{.3 \text{ meter}}$ or $\frac{.3 \text{ meter}}{1 \text{ foot}} = 1$

We use the one with *meter* on top.

$$4 \text{ feet} \times \frac{.3 \text{ meter}}{1 \text{ foot}} = \frac{\overset{1}{4 \cancel{\text{feet}}}}{1} \times \frac{.3 \text{ meter}}{\underset{1}{1 \cancel{\text{foot}}}} = 4 \times .3 \text{ meter} = 1.2 \text{ meters.}$$

Example 2: Change 19 cm. to inches.

Method I: We are changing from Metric to English. We use the left-hand chart.

1 cm. = .39 in. We replace the cm. with .39 in.

19 cm. = 19 × (.39 in.) = 7.41 in.

Method II: We use dimensional algebra. We use the conversion factor:

$$\frac{.39 \text{ in.}}{1 \text{ cm.}} = 1.$$

$$\frac{19 \text{ cm.}}{1} = \frac{19 \cancel{\text{cm.}}}{1} \times \frac{.39 \text{ in.}}{1 \cancel{\text{cm.}}} = 19 \times .39 \text{ in.} = 7.41 \text{ in.}$$

NOTE: The dimensions "cm." cancelled out.

Example 3: Change 5680 ml. to ounces.

Method I: In looking at our Metric to English chart there is no direct one step from milliliters to ounces.

First we can change ml. to liters. Then we change liters to quarts. Then we change quarts to ounces.

5680 ml. = 5.68 l.

Then we use the fact that 1 liter is 1.06 quarts.

5.68 l. = 5.68 × (1.06 qt.) = 6.0208 qt.

Then we convert 1 qt. to 32 oz.

C.0208 qt. = 6.0208 × (32 oz.) = 192.6656 oz.

We can round our answer off to one decimal place: 192.7 oz.

Method II: We use dimensional algebra. We need these conversion factors.

$$\frac{1 \text{ liter}}{1000 \text{ ml.}} \qquad \frac{1.06 \text{ qt.}}{1 \text{ liter}} \qquad \frac{32 \text{ oz.}}{1 \text{ qt.}}$$

Notice that reading from bottom of first fraction to top of last fraction we are doing: millimeter to liter to liter to quart to quart to ounces.

$$\frac{\text{liter}}{\text{ml.}} \longrightarrow \frac{\text{qt.}}{\text{liter}} \longrightarrow \frac{\text{oz.}}{\text{qt.}}$$

Here goes the solution:

$$\frac{5680 \cancel{\text{ml.}}}{1} \times \frac{\overset{1}{1 \cancel{\text{l.}}}}{1000 \cancel{\text{ml.}}} \times \frac{1.06 \cancel{\text{qt.}}}{\underset{1}{1 \cancel{\text{l.}}}} \times \frac{32 \text{ oz.}}{1 \cancel{\text{qt.}}} = \frac{5680}{1000} \times 1.06 \times 32 \text{ oz.}$$

$$= 192.6656 \text{ oz.}$$

Example 4: How many kilograms would a 120-lb. woman weigh?

Method I: Everywhere we see the word "pound" we replace it with "454 grams". Then we change grams to kilograms.

120 lb. = 120 × (454 grams) = 54,480 g. = 54.480 kg. or 54.48 kg.

Method II: We use dimensional algebra and the conversion factor $\frac{454 \text{ g.}}{1 \text{ lb.}}$

$$120 \text{ lb.} = \frac{\overset{1}{120 \cancel{\text{lb.}}}}{1} \times \frac{454 \text{ g.}}{\underset{1}{1 \cancel{\text{lb.}}}} = 54{,}480 \text{ g.} = 54.48 \text{ kg.}$$

Example 5: A small plane weighs 2 tons, while a helicopter weighs 1800 kg. Which is heavier, and by how much?

Method I: Let's change tons to kilograms by replacing ton with 2000 pounds and pounds with .454 kg.

$$2 \text{ tons} = 2 \times (2000 \text{ lb.}) = 4000 \text{ lb.} = 4000 \times (.454 \text{ kg.}) = 1816 \text{ kg.}$$

The small plane is heavier by 16 kg. (1816 − 1800 = 16)

Method II: Using dimensional algebra we will change kg. to lb.

We will use the conversion factor $\frac{2.2 \text{ lb.}}{1 \text{ kg.}}$

$$1800 \text{ kg.} = \frac{1800 \cancel{\text{kg.}}}{1} \times \frac{2.2 \text{ lb.}}{1 \cancel{\text{kg.}}} = 3960 \text{ lb.}$$

The helicopter weighs approximately 3960 lb.

The small plane weighs 2 tons or 4000 lb.

The small plane is heavier by approximately 40 pounds.

(NOTE: Changing 16 kg. to lbs. we get 16 kg. = 16 × 2.2 lb. = 35 lb. not 40 lbs. There is a discrepancy in both answers since the two systems' values (Metric and English) are not exactly interchangeable. The equivalences have been rounded off. That accounts for the 5 pound difference.)

The following is a more difficult example:

Example 6: Change 13 sq. ft. to sq. meters.

Method II: First, we will change the sq. ft. to sq. in.

$$13 \text{ sq. ft.} = 13 (\text{ft.} \times \text{ft.}) = 13 \times (12 \text{ in.} \times 12 \text{ in.}) = 1872 \text{ sq. in.}$$

Now we convert sq. in. to sq. meters by using the conversion factor:

$$\frac{1 \text{ meter}}{39.37 \text{ in.}}$$

$$1872 \text{ sq. in.} = 1872 \text{ in.} \times \text{in.} = \frac{1872 \text{ in.}}{1} \times \frac{1 \text{ in.}}{1} \times \frac{1 \text{ meter}}{39.37 \text{ in.}} \times \frac{1 \text{ meter}}{39.37 \text{ in.}}$$

$$= \frac{1872 \cancel{\text{in.}}}{1} \times \frac{\cancel{\text{in}}}{1} \times \frac{1 \text{ meter}}{39.37 \cancel{\text{in.}}} \times \frac{1 \text{ meter}}{39.37 \cancel{\text{in.}}} =$$

$$\frac{1872}{39.37 \times 39.37} = 1.207744351 \text{ m}^2 \text{ or } 1.2 \text{ sq. m.}$$

Here is the problem done in one long step from the beginning:

$$13 \text{ ft.}^2 = 13 \text{ ft.} \times \text{ft.} = \frac{13 \overset{1}{\cancel{\text{ft.}}} \times \overset{1}{\cancel{\text{ft.}}}}{1} \times \frac{12 \overset{1}{\cancel{\text{in.}}}}{1 \underset{1}{\cancel{\text{ft.}}}} \times \frac{12 \overset{1}{\cancel{\text{in.}}}}{1 \underset{1}{\cancel{\text{ft.}}}} \times \frac{1 \text{ m.}}{39.37 \underset{1}{\cancel{\text{in.}}}} \times \frac{1 \text{ m.}}{39.37 \underset{1}{\cancel{\text{in.}}}}$$

$$= \frac{13\ (12)\ (12)}{(39.37)\ (39.37)} \text{ meter} \times \text{meter} \approx 1.2 \text{ m}^2$$

NOTE: 1.2 m^2 means 1.2 × (meter) × (meter). Don't confuse with $(1.2 \text{ meter})^2$ which is (1.2 meter) × (1.2 meter) which equals 1.44 m^2.

CONVERSIONS BETWEEN THE ENGLISH AND METRIC SYSTEMS EXERCISES:

Use the entry from the appropriate conversion table. It is helpful to use your calculator for these exercises:

1) You are bottling your homemade wine, but you only have cups (8 ounces) to use to fill the bottles. Your bottles are $\frac{4}{5}$ liter bottles.

   a) How many 8-ounce cups in a liter?

   b) How many cups are needed to fill each bottle?

2) In New York City the average city block is 50 meters long, running from North to South.

   a) How many feet to an average city block?

   b) If your pace is 18 inches, how many paces would you need to run the entire block?

3) The distance between New York and Los Angeles is 4690 km. Your car registers mileage in miles rather than kilometers. If you are driving at the speed of 60 miles per hour, how many hours of driving before you arrive in New York (assuming no stops and starting from LA)?

4) A hectare is 2.5 times the size of an acre. Use this information to calculate the number of square feet in an acre.

5) The World Trade Center is 412 meters tall. The Empire State Building is 381 meters tall. How much taller is the World Trade Center? Give your answer in feet and inches (rounded off to nearest inch).

6) The heart is a muscular pump weighing about one-third of a kilogram. Convert this to pounds and ounces (rounded off to the nearest ounce).

7) Which is larger, a yard or a meter? And by how much?

8) The mass of the Earth is 6.588×10^{21} tons. Convert this to metric tons (1 metric ton = 1000 kg.)

9) Our windpipe, otherwise known as our trachea, is a cylindrical tube that conducts air to our lungs. It is cylindrical in shape with a length of 25 cm. and diameter of 25 mm.

   a) Find the length in inches.

   b) Find the diameter in inches.

   c) Find the circumference (in inches) using the formula ($C = \pi D$).

10) The average adult brain of a woman weighs 1250 grams. The average adult brain of a man weighs 1380 grams. Convert each of these to pounds and ounces (round off to nearest ounce).

Female ____________ Male ____________

11) The heaviest human being on record weighed 1070 pounds. Find his weight in kilograms.

12) The equator is 25,000 miles long. Find this length in meters.

13) The speed at which sound travels is greatly affected by temperature. At 20 degrees Celsius, the speed of sound is 344 meters per second. At 100 degrees Celsius, the speed is 386 meters per second. To get a feeling for how fast this is, convert all metric units to our comparable units.

14) The smallest known insects in the world are the "hairy winged beetles". They measure .008 inches in length. How many of these beetles could we put in a line to fill a meter distance?

15) A "kip" is 1000 pounds. Change 10 kips to kilograms.

16) The average quantity of sweat secreted in 24 hours varies from 700 grams to 900 grams. If a person secretes 750 grams one day, how many 8 ounce cups would that fill? (Round off to 1 decimal place.)

17) In the course of a day we take in about two and a half liters of water in food and drink. We lose one and a half liters as urine, and half a liter in perspiration. Almost half a liter is lost as vapor from our lungs. A negligible amount is lost in feces. To get an idea of how much this is, convert all the above measures into ounces. (Round to nearest ounce.)

2½ liters ____________

1½ liters ____________

1/2 liter ____________

18) A sprinter runs 100 meters. How many feet is this?

METRIC SYSTEM

Here is a checklist of words that are used in dealing with the metric system. See if you can identify each one and define them to yourself.

linear measure

weight

volume

capacity

meter

gram

liter

milli

centi

deci

deka

hecto

kilo

dimensional algebra

ANSWERS

PLACE VALUE AND LARGE NUMBERS
ANSWERS, page 5

1) a) 10 billions
   b) billions
   c) millions
   d) 100 millions

2) a) 10 trillion
   b) 100 billion
   c) 100 million
   d) Ones
   e) 10 billion
   f) thousands
   g) million

3) 28 million, 400 thousand, 13
4) 1 trillion, 1
5) 7 million, 489 thousand, 116
6) 40 million, 85 thousand, 111
7) 1 quintillion, 111 quadrillion, 111 trillion, 111 billion, 111 million, 111 thousand, 11
8) 1 quadrillion, 10 trillion, 101 billion, 10 million, 101 thousand, 10
9) 9 trillion, 999 billion, 909 million, 909 thousand, 99
10) 830 million, 38
11) 18 trillion, 300 billion, 38 thousand, 83
12) 147 thousand, 95
13) 18,034
14) 678,345,011
15) 2,000,000,345,405
16) 19,000,000,475,238,000
17) 93,432,780,215
18) 1,000,000,000,000,000,005
19) 13,013,013
20) 999,099
21) 247,812
22) 3,000,004,000,003,000,000,007,000
23) 3,004,899
24) 4,000,000,500,020
25) 921,099
26) 45,000,000,000,000,000,000,000,000,000
27) 250,000,000 (or $\frac{1}{4}$ of 1,000,000,000 or $\frac{1000000000}{4}$)

EXPONENTS
ANSWERS, page 12

1) 3^5
2) 6^3
3) 8^9
4) 9^8
5) 10^4
6) $(3^2)^5$
7) $2 \times 2 \times 2 \times 2 \times 2 \times 2 \times 2 = 128$
8) $3 \times 3 = 9$
9) $4 \times 4 \times 4 = 64$
10) $5 \times 5 \times 5 = 125$
11) $7 \times 7 = 49$
12) $8 \times 8 \times 8 = 512$
13) $11 \times 11 = 121$
14) $9 \times 9 \times 9 = 729$
15) $2 \times 2 \times 2 \times 2 \times 2 = 32$
16) $3 \times 3 \times 3 \times 3 \times 3 = 243$
17) $10^3 = 10 \times 10 \times 10 = 1000$
    $10^4 = 10 \times 10 \times 10 \times 10 = 10{,}000$
    $10^5 = 10 \times 10 \times 10 \times 10 \times 10 = 100{,}000$
    $10^6 = 10 \times 10 \times 10 \times 10 \times 10 \times 10 = 1{,}000{,}000$
    The exponent tells how many *zeros* in the number.
18) $2^8 = 256$
    $2^7 = 128$
    $2^6 = 64$
    $2^5 = 32$
    $2^4 = 16$
    $2^3 = 8$
    $2^2 = 4$
    $2^1 = 2$
    All the values are *halved* as the exponent decreases.
    $2^0 = 1$
19) 4,782,969
20) 16,777,216
21) 161,051
22) 48,828,125
23) 33,554,432
24) 1,030,301
25) 11,390,625
26) 4,004,001
27) 704,969
28) 3,944,196
29) 4^5
30) 3^{13}

(Answers continued on next page)

31) 2^{13}
32) 10^7
33) 12^{20}
34) 2^{325}
35) $9^{\square+\triangle}$
36) $3^{3^3} = 3^{3\times3\times3} = 3^{27}$
37) $10^{10^{10}} = 10^{10 \text{ billion}}$
Write a 1 with 10 billion zeros after it.
38) a) $2^2 = 4$
b) $2^3 = 8$
c) $2^4 = 16$
d) $2^{10} = 1024$
39) a) $81 = 3^4$
b) $3^7 = 2187$
40) 60 days total (1 more day)

SHORTCUTS TO MENTAL ARITHMETIC ANSWERS, page 17

1) 40,960
2) 1270
3) 720
4) 10,040
5) 52,280
6) 43,810
7) 206,110
8) 140
9) 16,650
10) 340,020
11) 960
12) 230
13) 48,830
14) 4380
15) 4990
16) 204,800
17) 4,002,900
18) 1200
19) 3200
20) 24,100
21) 100,500
22) 532,100
23) 19,700
24) 1,999,900
25) 2,148,900
26) 192
27) 342
28) 361
29) 224
30) 195
31) 255
32) 256
33) 187
34) 216
35) 209

RULES FOR DIVISIBILITY EXERCISES ANSWERS, page 21

1) divisible by 3, 5, and 9
2) divisible by 2, 4, and 8
3) divisible by 3, 5, and 9
4) divisible by 3 and 9
5) 2, 3, 4, 6, and 8
6) 3, 9
7) 3
8) 2, 3, 4, 6, 8
9) 3
10) 3
11) 2, 3, 6, 9
12) 2, 3, 6, 9
13) 2, 4, 8
14) 2, 3, 4, 6, 8, 9
15) 2, 3, 5, 6, 9, 10
16) 5
17) 2, 3, 4, 6, 8
18) 3
19) 2, 3, 6, 9
20) 2, 3, 6, 9
21) Every third consecutive number is a multiple of 3. Every other number is a multiple of 2 (even). In three consecutive numbers there is at least one number that is even and one that is divisible by 3. When we multiply, we get a number divisible by 2 × 3 which is 6.
22) a) 7 b) 7 c) 0 or 9
23) a) 4 b) 1, 4, 7 c) 1, 4, 7

PRIME FACTORIZATION OF WHOLE NUMBERS ANSWERS, page 27

1) Say you want to cross out multiples of 3, you start with the number 3 and cross out every *third* number *after* 3.

(Answers continued on next page)

2) 2, 3, 5, 7, 11, 13, 17, 19, 23, 29, 31, 37, 41, 43, 47, 53, 59, 61, 67, 71, 73, 79, 83, 89, 97
3) $100 = 2 \times 50 = 4 \times 25 = 10 \times 10$
4) Any even number is divisible by 2 so it can be written as $2 \times \square$
5) $12 = 2^2 \times 3$
6) $40 = 2^3 \times 5$
7) $36 = 2^2 \times 3^2$
8) $50 = 2 \times 5^2$
9) $81 = 3^4$
10) $242 = 2 \times 11^2$
11) $65 = 5 \times 13$
12) $459 = 3^3 \times 17$
13) $111 = 3 \times 37$
14) $2001 = 3 \times 23 \times 29$
15) $360 = 2^3 \times 3^2 \times 5$
16) $1000 = 2^3 \times 5^3$
17) $48 = 2^4 \times 3$
18) $128 = 2^7$
19) $92 = 2^2 \times 23$
20) $60 = 2^2 \times 3 \times 5$
21) $216 = 2^3 \times 3^3$
22) $336 = 2^4 \times 3 \times 7$
23) $125 = 5^3$
24) $27 = 3^3$
25) See Table of Prime Numbers (See page 325)

PICTORIAL PATTERNS USING PRIME FACTORIZATIONS
ANSWERS, page 31

1) 13

2) $9 = 3^2$

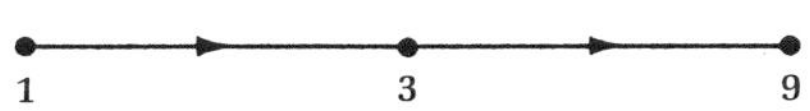

3) $22 = 2 \times 11$

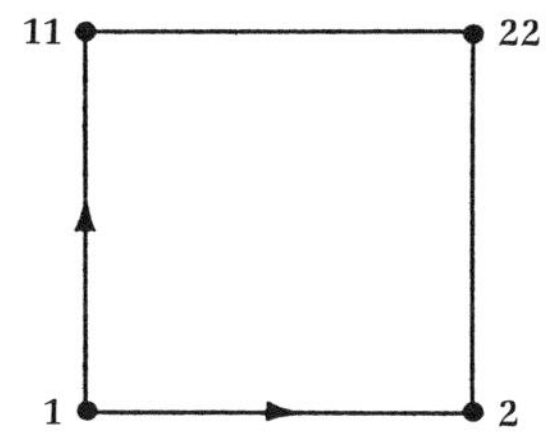

4) $45 = 3^2 \times 5$

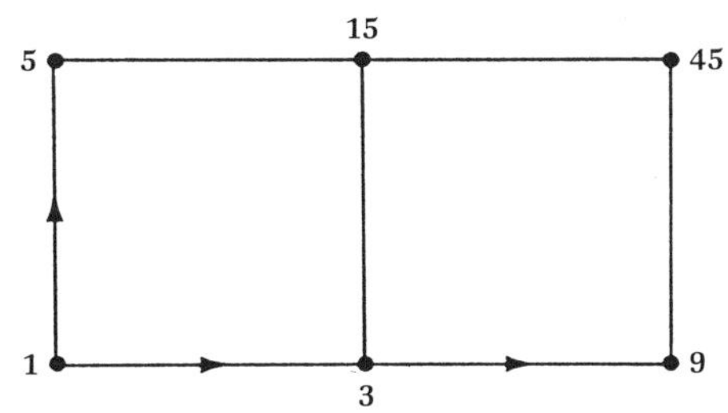

5) $75 = 3 \times 5^2$

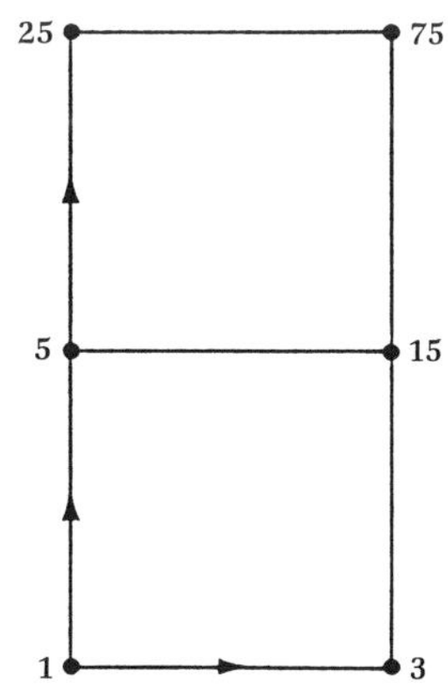

6) $24 = 2^3 \times 3$

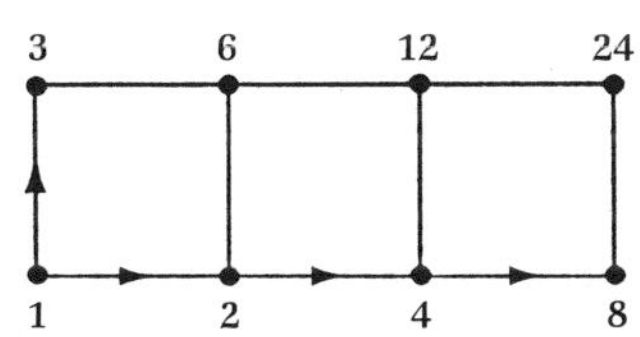

(Answers continued on next page)

7) $100 = 2^2 \times 5^2$

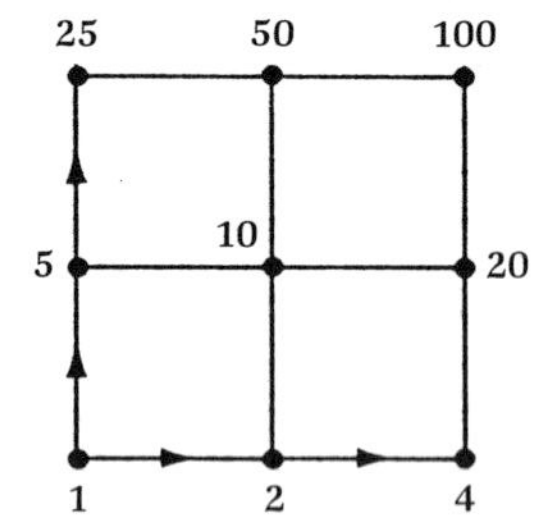

8) $42 = 2 \times 3 \times 7$

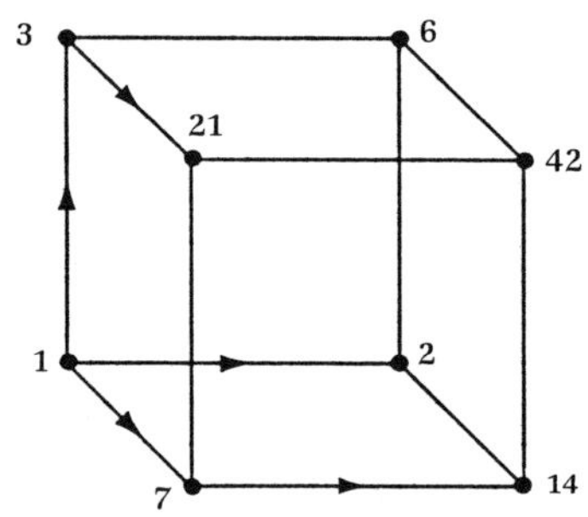

9) $54 = 2 \times 3^3$

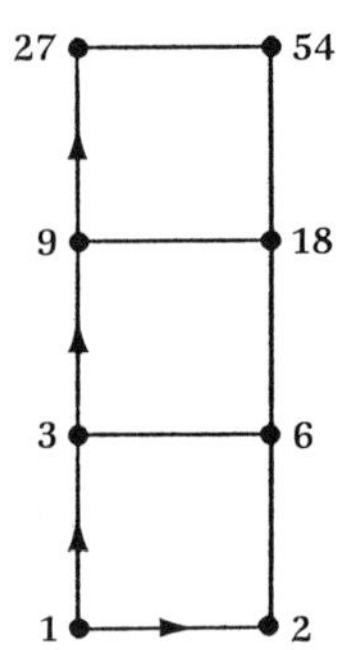

10) $60 = 2^2 \times 3 \times 5$

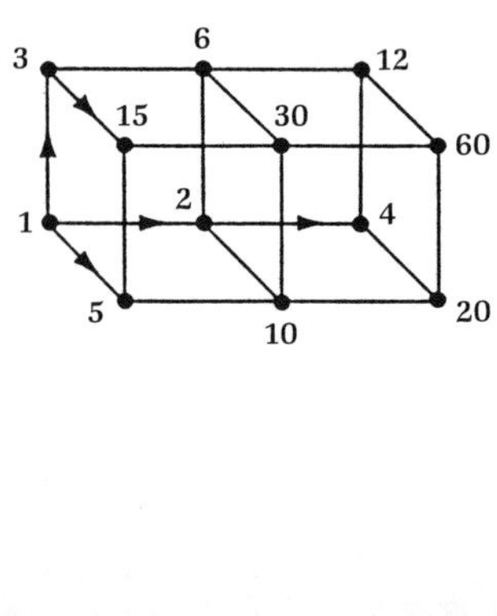

11) a = 1 b = 2 c = 4
d = 5 e = 10 f = 20
g = 40 h = 50 i = 100
j = 200
200 is the number

12) a = 1 b = 2 c = 8
d = 16
16 is the number

13) a = 1 b = 3 c = 9
d = 27 e = 54 f = 18
g = 2 h = 4 i = 36
j = 108
108 is the number

14) a = 1 b = 2 c = 4
d = 8 e = 16 f = 80
g = 40 h = 5
80 is the number

GREATEST COMMON DIVISOR
ANSWERS, page 35

1) 6
2) 3
3) 11
4) 10
5) 20
6) 18
7) 3
8) 10
9) 10
10) 1
11) 3
12) 6
13) 72
14) 1
15) 7
16) 11
17) 25
18) $2^3 \times 3 = 24$

LEAST COMMON MULTIPLE
ANSWERS, page 39

1) $10 = 2 \times 5$
$12 = 2 \times 2 \times 3$
L.C.M. $= 2 \times 2 \times 3 \times 5 = 60$

(Answers continued on next page)

2) $4 = 2^2$
$6 = 2 \times 3$
$5 = 5$
L.C.M. $= 2^2 \times 3 \times 5 = 60$

3) $3 = 3$
$2 = 2$
$5 = 5$
L.C.M. $= 2 \times 3 \times 5 = 30$

4) $100 = 2^2 \times 5^2$
$25 = 5^2$
L.C.M. $= 2^2 \times 5^2 = 100$

5) $18 = 2 \times 3^2$
$21 = 3 \times 7$
L.C.M. $= 2 \times 3^2 \times 7 = 126$

6) $12 = 2^2 \times 3$
$15 = 3 \times 5$
$20 = 2^2 \times 5$
L.C.M. $= 2^2 \times 3 \times 5 = 60$

7) $27 = 3^3$
$18 = 2 \times 3^2$
$36 = 2^2 \times 3^2$
$4 = 2^2$
L.C.M. $= 2^2 \times 3^3 = 108$

8) $6 = 2 \times 3$
$15 = 3 \times 5$
$10 = 2 \times 5$
L.C.M. $= 2 \times 3 \times 5 = 30$

9) $3 = 3$
$4 = 2 \times 2$
$8 = 2 \times 2 \times 2$
$6 = 2 \times 3$
L.C.M. $= 2^3 \times 3 = 24$

10) $10 = 10$
$100 = 10^2$
$1000 = 10^3$
L.C.M. $= 10^3 = 1000$

11) $60 = 2^2 \times 3 \times 5$
$72 = 2^3 \times 3^2$
L.C.M. $= 2^3 \times 3^2 \times 5 = 360$

12) $8 = 2^3$
$9 = 3^2$
$12 = 2^2 \times 3$
L.C.M. $= 2^3 \times 3^2 = 72$

13) $12 = 2^2 \times 3$
$16 = 2^4$
L.C.M. $= 2^4 \times 3 = 48$

14) $10 = 2 \times 5$
$8 = 2^3$
L.C.M. $= 2^3 \times 5 = 40$

15) $24 = 2^3 \times 3$
$15 = 3 \times 5$
L.C.M. $= 2^3 \times 3 \times 5 = 120$

16) $35 = 5 \times 7$
$315 = 3^2 \times 5 \times 7$
L.C.M. $= 3^2 \times 5 \times 7 = 315$

17) $48 = 2^4 \times 3$
$192 = 2^6 \times 3$
L.C.M. $= 2^6 \times 3 = 192$

18) $65 = 5 \times 13$
$91 = 7 \times 13$
L.C.M. $= 5 \times 7 \times 13 = 455$

19) $11 = 11$
$13 = 13$
$7 = 7$
L.C.M. $= 7 \times 11 \times 13 = 1001$

20) $485 = 5 \times 97$
$291 = 3 \times 97$
L.C.M. $= 3 \times 5 \times 97 = 1455$

21) $6 = 2 \times 3$
$8 = 2^3$
L.C.M. $= 2^3 \times 3 = 24$
(24 minutes later)

22) 8:25 p.m.

23) 150 boxes

24) 240 minutes later

25) 120 minutes (2 hours)

26) 8 gallons in 18 mpg car
9 gallons in 16 mpg car

TWO INTERESTING WAYS TO GET LCMs ANSWERS, page 45

1) 36
2) 30
3) 189
4) 54
5) 336
6) 90
7) 1000
8) 144

(Answers continued on next page)

9) 150
10) 80
11) 420
12) 200
13) 70
14) 60
15) 60
16) 72
17) 48
18) 450
19) 639
20) 288
21) 455
22) 1716
23) 1000
24) 84

FRACTIONS: A CONCRETE APPROACH
ANSWERS, page 51

1) $3\frac{1}{4}$
2) $3\frac{4}{7}$
3) $3\frac{3}{5}$
4) $3\frac{1}{6}$
5) $13\frac{1}{2}$
6) $2\frac{4}{7}$
7) $11\frac{1}{9}$
8) $2\frac{1}{5}$
9) $500\frac{1}{2}$
10) $6\frac{3}{7}$
11) $8\frac{9}{10}$
12) $2\frac{11}{12}$
13) $7\frac{1}{5}$
14) $49\frac{1}{26}$
15) $9\frac{5}{20} = 9\frac{1}{4}$
16) 14
17) $1\frac{2}{15}$
18) $8\frac{10}{11}$
19) $13\frac{2}{3}$
20) $66\frac{2}{3}$
21) $1\frac{49}{126}$
22) $5\frac{3}{10}$
23) $12\frac{1}{8}$
24) $33\frac{1}{3}$
25) $37\frac{1}{2}$
26) $\frac{43}{4}$
27) $\frac{129}{10}$
28) $\frac{42}{5}$
29) $\frac{7}{4}$
30) $\frac{5}{2}$
31) $\frac{63}{4}$
32) $\frac{11}{3}$
33) $\frac{23}{2}$
34) $\frac{97}{8}$
35) $\frac{63}{8}$
36) $\frac{29}{3}$
37) $\frac{89}{3}$
38) $\frac{100}{3}$
39) $\frac{200}{3}$
40) $\frac{25}{2}$
41) $\frac{31}{6}$
42) $\frac{375}{8}$
43) $\frac{415}{8}$
44) $\frac{31}{16}$
45) $\frac{999}{10}$
46) $\frac{11}{4}$
47) $\frac{53}{6}$
48) $\frac{1223}{12}$
49) $\frac{3}{2}$
50) $\frac{379}{19}$

ANSWERS, page 52

Shade each figure to represent the given fraction:

51) $\frac{1}{2}$

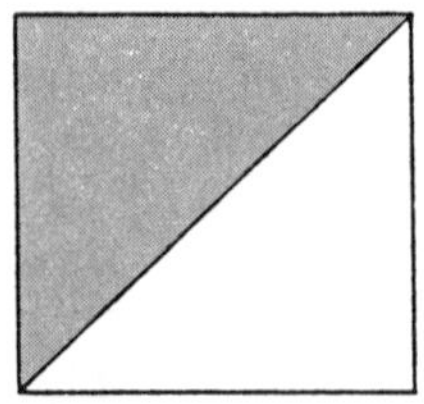

52) $\frac{4}{6}$

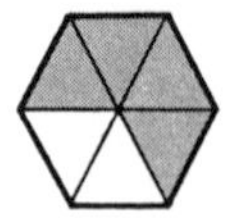

53) $\frac{3}{6}$

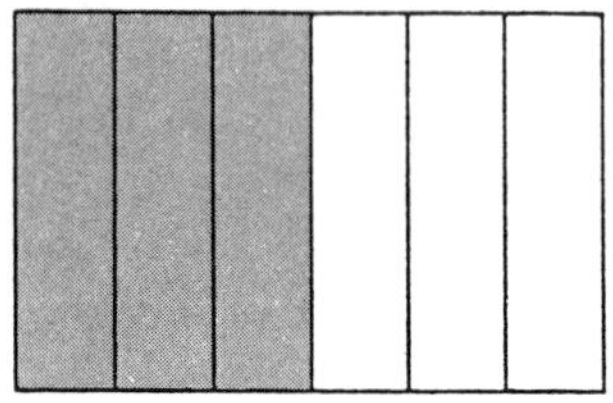

54) $\frac{2}{3}$

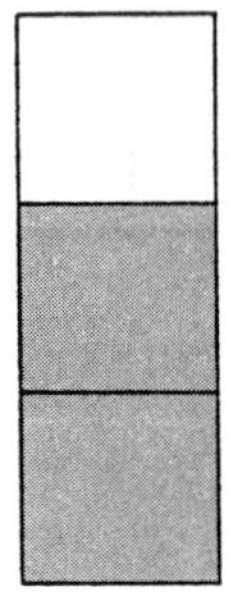

55) $\frac{1}{2}$

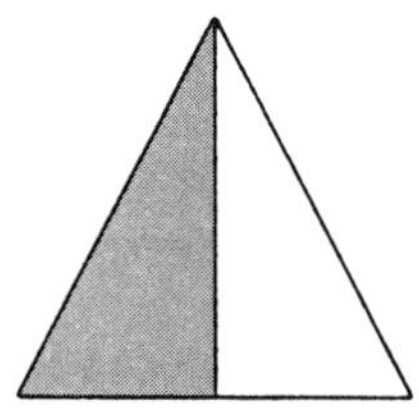

56) $\frac{3}{4}$

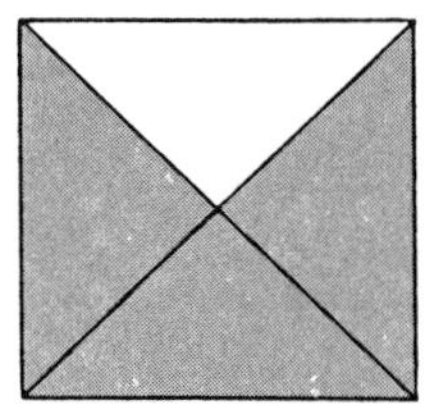

57) Shade $\frac{3}{5}$ of the total region encompassed by these 5 hexagons:

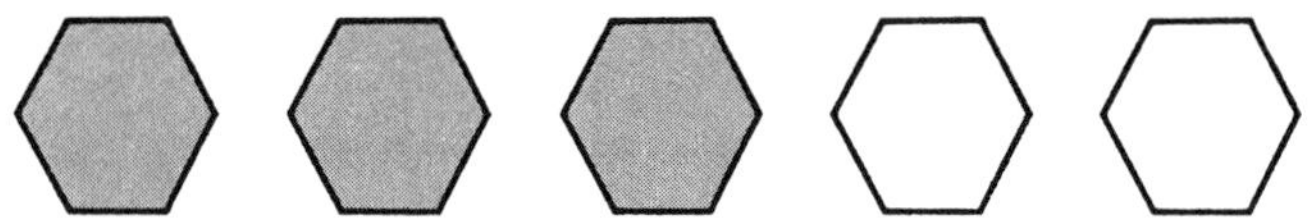

58) Mark one-tenth of the total area in these 5 circles:

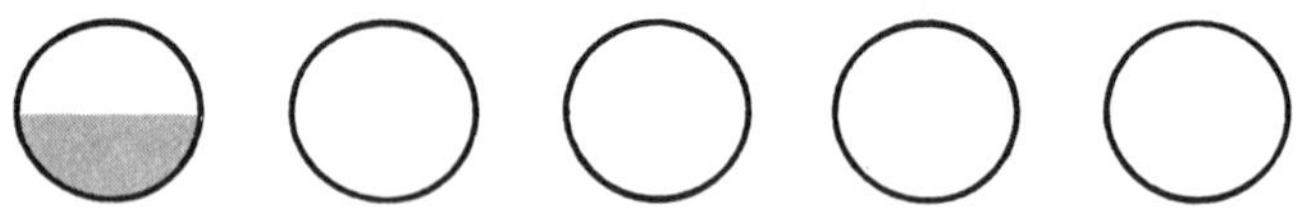

59) Shade $\frac{2}{3}$ of the total area encompassed by these 6 squares:

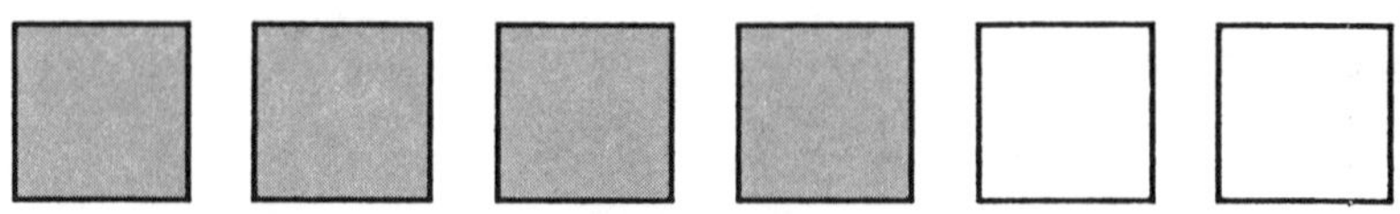

ANSWERS, page 53

60) $\frac{1}{4}$

61) $\frac{3}{8}$

62) $\frac{1}{3}$

63) $\frac{3}{4}$

64) $\frac{3}{5}$

65) $\frac{3}{24} = \frac{1}{8}$

66) $\frac{1}{4}$

67) $\frac{1}{64}$

68) $\frac{1}{12}$

69)  $= \frac{2}{3}$,  $= \frac{2}{5}$, both kinds of lines $= \frac{4}{15} = (\frac{2}{3} \times \frac{2}{5})$

EQUIVALENT FRACTIONS

ANSWERS, page 59

1) $\frac{1}{3} = \frac{2}{6} = \frac{3}{9} = \frac{4}{12} = \frac{5}{15} = \frac{6}{18} = \ldots$

2) $\frac{2}{5} = \frac{4}{10} = \frac{6}{15} = \frac{8}{20} = \frac{10}{25} = \frac{12}{30} = \ldots$

3) $\frac{5}{3} = \frac{10}{6} = \frac{15}{9} = \frac{20}{12} = \frac{25}{15} = \frac{30}{18} = \ldots$

4) $\frac{3}{2} = \frac{6}{4} = \frac{9}{6} = \frac{12}{8} = \frac{15}{10} = \frac{18}{12} = \ldots$

5) $\frac{3}{4} = \frac{6}{8} = \frac{9}{12} = \frac{12}{16} = \frac{15}{20} = \frac{18}{24} = \ldots$

6) $\frac{7}{8} = \frac{14}{16} = \frac{21}{24} = \frac{28}{32} = \frac{35}{40} = \frac{42}{48} = \ldots$

7) $\frac{1}{2} = \frac{2}{4} = \frac{3}{6} = \frac{4}{8} = \frac{5}{10} = \frac{6}{12} = \ldots$

8) $\frac{3}{10} = \frac{6}{20} = \frac{9}{30} = \frac{12}{40} = \frac{15}{50} = \frac{18}{60} = \ldots$

9) $\frac{1}{8} = \frac{2}{16} = \frac{3}{24} = \frac{4}{32} = \frac{5}{40} = \frac{6}{48} = \ldots$

10) $\frac{1}{6} = \frac{2}{12} = \frac{3}{18} = \frac{4}{24} = \frac{5}{30} = \frac{6}{36} = \ldots$

11) $\frac{5}{6} = \frac{10}{12} = \frac{15}{18} = \frac{20}{24} = \frac{25}{30} = \frac{30}{36} = \ldots$

12) $\frac{8}{15} = \frac{16}{30} = \frac{24}{45} = \frac{32}{60} = \frac{40}{75} = \frac{48}{90} = \ldots$

13) $\frac{9}{10} = \frac{18}{20} = \frac{27}{30} = \frac{36}{40} = \frac{45}{50} = \frac{54}{60} = \ldots$

14) $\frac{2}{3} = \frac{4}{6} = \frac{6}{9} = \frac{8}{12} = \frac{10}{15} = \frac{12}{18} = \ldots$

15) $\frac{7}{9} = \frac{14}{18} = \frac{21}{27} = \frac{28}{36} = \frac{35}{42} = \frac{42}{54} = \ldots$

16) $\frac{3}{4} = \frac{6}{8}$

17) $\frac{4}{5} = \frac{20}{25}$

18) $\frac{5}{6} = \frac{20}{24}$

19) $\frac{1}{2} = \frac{49}{98}$

20) $\frac{3}{5} = \frac{51}{85}$

21) $\frac{7}{9} = \frac{77}{99}$

22) $\frac{0}{8} = \frac{0}{16}$

(Answers continued on page 304)

23) $\frac{0}{8} = \frac{0}{9}$

24) $\frac{7}{5} = \frac{77}{55}$

25) $\frac{2}{3} = \frac{26}{39}$

26) $\frac{5}{8} = \frac{625}{1000}$

27) $\frac{2}{5} = \frac{24}{60}$

28) $\frac{8}{5} = \frac{24}{15}$

29) $1\frac{1}{2} = \frac{3}{2} = \frac{6}{4}$

30) $3\frac{3}{5} = \frac{18}{5} = \frac{36}{10}$

31) $15\frac{1}{3} = \frac{46}{3} = \frac{184}{12}$

32) $8\frac{4}{5} = \frac{44}{5} = \frac{88}{10}$

33) $2\frac{2}{3} = \frac{8}{3} = \frac{40}{15}$

34) $6\frac{7}{8} = \frac{55}{8} = \frac{220}{32}$

35) $4\frac{2}{3} = \frac{14}{3} = \frac{42}{9}$

36) $\frac{4}{5}$

37) $\frac{2}{3}$

38) $\frac{2}{7}$

39) $\frac{1}{2}$

40) $\frac{3}{8}$

41) $\frac{2}{3}$

42) $\frac{8}{9}$

43) $\frac{1}{4}$

44) $\frac{9}{11}$

45) $\frac{3}{4}$

46) $\frac{7}{12}$

47) $\frac{3}{5}$

48) $\frac{5}{2}$

49) $\frac{4}{3}$

50) 8

51) $\frac{7}{12}$

52) $\frac{11}{16}$

53) already reduced

54) $\frac{37}{4}$

55) $\frac{3}{25}$

56) $\frac{3}{7}$

57) already reduced

58) 0

59) $\frac{7}{10}$

60) $\frac{17}{24}$

61) T

62) T

63) F $\frac{5}{8} = \frac{20}{32}$

64) F $\frac{12}{35} = \frac{24}{70}$

65) F $\frac{1}{3} = \frac{22}{66}$

66) T

67) F $\frac{5}{4} = \frac{125}{100}$

68) F $\frac{3}{2} = \frac{54}{36}$

69) F $\frac{24}{5} = \frac{72}{15}$

70) $\frac{5}{6} = \frac{10}{12} = \frac{15}{18} = \frac{40}{48}$

GRAPHING FRACTIONS AND ORDERING FRACTIONS ANSWERS, page 67

1) 

2) 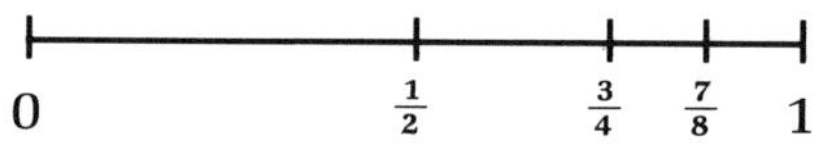

3) 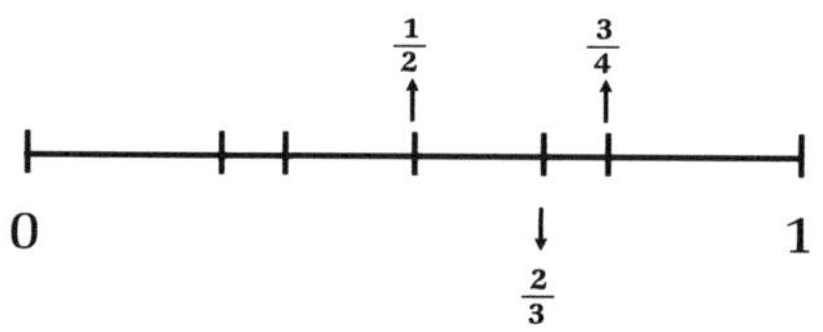

4) 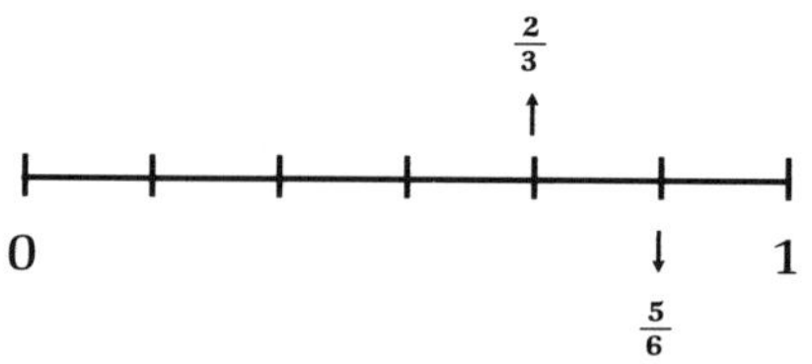

5) 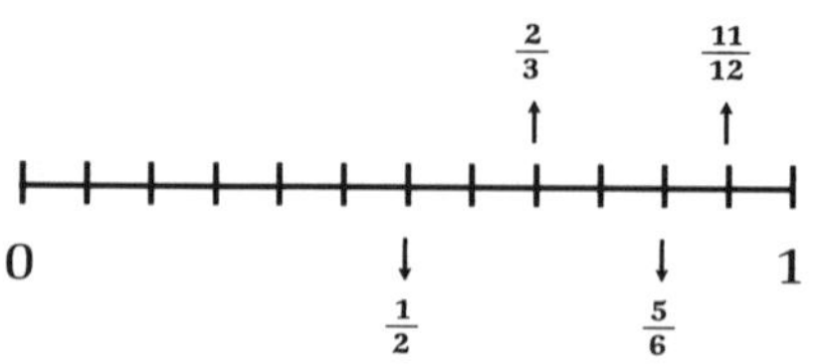

$\frac{1}{2} = \frac{6}{12}$ $\frac{5}{6} = \frac{10}{12}$

$\frac{2}{3} = \frac{8}{12}$ $\frac{11}{12} = \frac{11}{12}$

6) 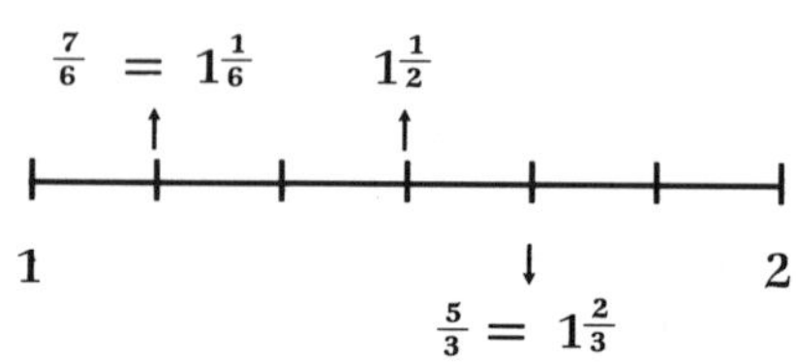

(Answers continued on next page)

7) 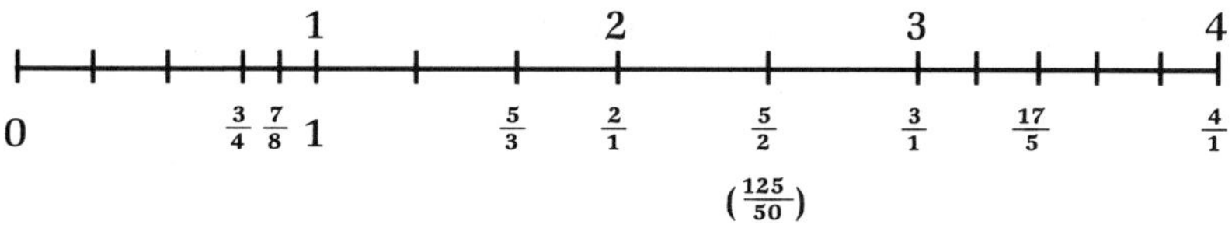

8) between 3 and 4, closer to 4
9) between 4 and 5, closer to 4
10) between 1 and 2, closer to 1
11) between 0 and 1, closer to 1
12) between 6 and 7, closer to 7
13) between 2 and 3, exactly midway
14) between 8 and 9, closer to 8
15) between 0 and 1, closer to 1
16) $\frac{2}{3}$, $\frac{5}{6}$, $\frac{7}{8}$, $\frac{11}{12}$
17) $\frac{2}{3}$, $\frac{7}{8}$, $\frac{13}{14}$, $\frac{15}{16}$
18) $\frac{9}{10}$, $\frac{9001}{10000}$, $\frac{901}{1000}$, $\frac{909}{1000}$, $\frac{99}{100}$
19) $\frac{5}{2}$, $\frac{28}{11}$, $\frac{19}{7}$
20) $\frac{13}{9}$, $\frac{3}{2}$, $\frac{17}{11}$, $\frac{14}{9}$, $\frac{11}{7}$
21) yes
22) yes
23) no
24) no
25) yes
26) no
27) no
28) yes
29) no
30) yes
31) yes
32) yes
33) no
34) yes
35) yes
36) no
37) yes
38) yes
39) no
40) yes
41) no
42) no
43) no
44) yes
45) yes
46) no
47) yes
48) yes
49) yes
50) no
51) $1\frac{1}{3}$, $1\frac{1}{4}$, $1\frac{1}{5}$, $1\frac{1}{6}$, $1\frac{1}{7}$, ... also $1\frac{2}{5}$, $1\frac{2}{7}$, $1\frac{3}{7}$, ... also $1\frac{4}{9}$, ...
52) $1\frac{3}{4}$, $1\frac{3}{5}$, $1\frac{5}{6}$, $1\frac{2}{3}$, $1\frac{6}{7}$, ... There are many more.
53) $2\frac{1}{2}$, $2\frac{5}{12}$, $2\frac{7}{12}$, $2\frac{3}{5}$, $2\frac{7}{10}$, etc.

ADDITION AND SUBTRACTION OF FRACTIONS ANSWERS, page 77

1) $\frac{2}{3}$
2) 1
3) $\frac{14}{19}$
4) $1\frac{1}{2} = \frac{3}{2}$
5) $1\frac{2}{3} = \frac{5}{3}$
6) $2\frac{1}{6} = \frac{13}{6}$
7) $1\frac{1}{4} = \frac{5}{4}$
8) $1\frac{1}{4} = \frac{5}{4}$
9) $1\frac{1}{6} = \frac{7}{6}$
10) $1\frac{17}{24} = \frac{41}{24}$
11) $2\frac{3}{4} = \frac{11}{4}$
12) $1\frac{11}{40} = \frac{51}{40}$
13) $1\frac{53}{72} = \frac{125}{72}$
14) $\frac{17}{22}$
15) 7
16) $5\frac{3}{10}$
17) $5\frac{3}{4}$
18) $11\frac{1}{4}$
19) $6\frac{3}{8}$
20) $16\frac{1}{6}$
21) $24\frac{5}{8}$
22) $8\frac{5}{12}$
23) $31\frac{17}{30}$
24) $39\frac{3}{10}$
25) $49\frac{5}{8}$
26) $\frac{1}{4}$
27) $\frac{4}{9}$
28) 1
29) $\frac{1}{6}$
30) $1\frac{1}{5} = \frac{6}{5}$
31) $\frac{1}{12}$
32) $\frac{1}{24}$
33) $\frac{1}{8}$
34) $\frac{1}{15}$
35) $9\frac{1}{2}$
36) $7\frac{1}{2}$
37) $2\frac{1}{2}$
38) $6\frac{1}{4}$
39) $5\frac{7}{8}$
40) $6\frac{3}{8}$
41) $9\frac{1}{6}$
42) $31\frac{17}{30}$
43) $1\frac{1}{2}$
44) $8\frac{5}{8}$
45) $7\frac{7}{10}$
46) $67\frac{2}{3}$
47) $4\frac{21}{40}$
48) $6\frac{7}{10}$
49) $2\frac{1}{2}$
50) $4\frac{9}{10}$
51) $\frac{7}{2}$
52) $\frac{23}{24}$
53) $6\frac{2}{3}$
54) $3\frac{5}{12}$
55) $\frac{1}{12}$
56) $\frac{1}{8}$

APPLICATIONS USING ADDITION AND SUBTRACTION OF FRACTIONS
ANSWERS, page 81

1) $17\frac{1}{2}$
2) 41
3) $39\frac{11}{20}$
4) $2\frac{1}{2}$
5) $\frac{9}{20}$ larger
6) $5\frac{19}{24}$ larger
7) $35\frac{7}{12}$
8) $5\frac{5}{8}$
9) $5\frac{7}{8}$
10) $32\frac{3}{8}$
11) $6\frac{31}{32}$
12) loss; down $\frac{1}{2}$; new price $9\frac{7}{8}$
13) $141\frac{3}{4}$
14) $69\frac{1}{4}$
15) A = $95\frac{1}{4}$"
B = $50\frac{7}{8}$"
C = $43\frac{5}{8}$"
D = 103"

SOME CHALLENGING FRACTION PROBLEMS AND PATTERNS
ANSWERS, page 85

1) a) $\frac{3}{4}$
b) $\frac{7}{8}$
c) $\frac{15}{16}$
d) $\frac{31}{32}$
e) $\frac{1}{2} + \frac{1}{4} + \frac{1}{8} + \frac{1}{16} + \frac{1}{32} + \frac{1}{64} = \frac{63}{64}$
$\frac{1}{2} + \frac{1}{4} + \frac{1}{8} + \frac{1}{16} + \frac{1}{32} + \frac{1}{64} + \frac{1}{128} = \frac{127}{128}$
$\frac{1}{2} + \frac{1}{4} + \frac{1}{8} + \frac{1}{16} + \frac{1}{32} + \frac{1}{64} + \frac{1}{128} + \frac{1}{256} = \frac{255}{256}$
f) The numerator is one less than the denominator.

2) a) $\frac{4}{9}$
b) $\frac{13}{27}$
c) $\frac{40}{81}$
d) $\frac{1}{3} + \frac{1}{9} + \frac{1}{27} + \frac{1}{81} + \frac{1}{243} = \frac{121}{243}$
$\frac{1}{3} + \frac{1}{9} + \frac{1}{27} + \frac{1}{81} + \frac{1}{243} + \frac{1}{729} = \frac{364}{729}$
e) Numerator is *half* of one less than the denominator, i.e.,

$$\frac{\frac{1}{2}(81 - 1)}{81} = \frac{\frac{1}{2}(80)}{81} = \frac{40}{81}$$

3) $5\frac{1}{2}$
4) 2 lbs.
5) $4 \times \frac{4}{5} = 4 - \frac{4}{5} = 3\frac{1}{5}$
a) $5 \times \frac{5}{6} = 5 - \frac{5}{6} = 4\frac{1}{6}$
$6 \times \frac{6}{7} = 6 - \frac{6}{7} = 5\frac{1}{7}$
$7 \times \frac{7}{8} = 7 - \frac{7}{8} = 6\frac{1}{8}$
$8 \times \frac{8}{9} = 8 - \frac{8}{9} = 7\frac{1}{9}$
etc.
b) $13 \times \frac{13}{14} = 13 - \frac{13}{14} = 12\frac{1}{14}$
c) $n \times \frac{n}{n+1} = n - \frac{n}{n+1} = (n - 1) + \frac{1}{n+1}$
d) If we added 14 fractions of the form $\frac{13}{14}$ we would get 13, since $14 \times \frac{13}{14} = 13$. Since we added 13 fractions instead of 14, we have $13 - \frac{13}{14}$ (one fraction less).
6) $\frac{1}{2} + \frac{2}{1} + \frac{1}{3} + \frac{3}{1} + \frac{2}{3} + \frac{3}{2} = 8$
7) 84 hours

MULTIPLICATION AND DIVISION OF FRACTIONS
ANSWERS, page 91

1) $\frac{2}{3}$

2) $\frac{3}{2}$

3) $\frac{8}{15}$

4) $\frac{3}{16}$

5) $\frac{12}{25}$

6) $\frac{5}{9}$

7) $\frac{7}{18}$

8) $\frac{3}{5}$

9) 9

10) 15

11) $\frac{1}{8}$

12) 16

13) $\frac{7}{6} = 1\frac{1}{6}$

14) $\frac{8}{3} = 2\frac{2}{3}$

15) $\frac{1}{8}$

16) $\frac{8}{3} = 2\frac{2}{3}$

17) 11

18) $\frac{1}{24}$

19) $\frac{13}{2} = 6\frac{1}{2}$

20) $\frac{25}{2} = 12\frac{1}{2}$

21) 200

22) $\frac{7}{8}$

23) $\frac{25}{9} = 2\frac{7}{9}$

24) $\frac{33}{8} = 4\frac{1}{8}$

25) $\frac{3}{4}$

26) $\frac{15}{4}$

27) $\frac{9}{4}$

28) $\frac{8}{27}$

29) 1

30) $\frac{9}{5} = 1\frac{4}{5}$

31) 8

32) 6

33) 22

34) $\frac{1}{8}$

35) 35

36) 6

37) 1

38) $\frac{9}{4}$

39) $\frac{1}{6}$

40) $\frac{21}{8} = 2\frac{5}{8}$

41) $\frac{7}{24}$

42) 15

43) 9

44) $\frac{1}{9}$

45) 4

46) $\frac{1}{4}$

47) 2

48) $\frac{1}{9}$

49) $\frac{4}{5}$

50) $\frac{7}{10}$

51) $\frac{2}{3}$

52) $\frac{1}{2}$

53) $\frac{3}{4}$

54) $\frac{3}{2}$

55) $\frac{1}{6}$

56) 12

57) 8

58) 3

59) $\frac{9}{2} = 4\frac{1}{2}$

60) $\frac{3}{5}$

61) $\frac{4}{3} = 1\frac{1}{3}$

62) $\frac{6}{5}$ or $1\frac{1}{5}$

63) $\frac{4}{1} = 4$

64) $\frac{2}{5}$

65) $\frac{10}{53}$

66) $\frac{19}{85}$

67) $\frac{1}{5}$

68) $\frac{1}{2}$

69) 10

70)* $\frac{2}{0}$ which is said to be "undefined"

71) 1

(Answers continued on next page)

72) $\frac{9}{16}$

73) $\frac{7}{5}$ or $1\frac{2}{5}$

74) $\frac{6}{5}$ or $1\frac{1}{5}$

75) $\frac{19}{18}$ or $1\frac{1}{18}$

APPLICATIONS OF MULTIPLICATION AND DIVISION OF FRACTIONS
ANSWERS, page 97

1) 24 questions
2) 55 miles
3) $33,372
4) 24 bags
5) $46\frac{1}{4}$ cm.
6) 24 pans
7) 20 mpg
8) 400 pens
9) 132 miles
10) 70 mph
11) 60 kg.
12) 18 lbs.
13) 76 gallons
14) $117\frac{1}{2}$ lbs.
15) $7\frac{1}{2}$ gallons
16) a) 9 ft.
    b) $8\frac{1}{10}$ ft.
    c) $7\frac{29}{100}$ feet
    d) 5 bounces
17) $1\frac{2}{5}$ lb.
18) 100
19) 2510

MORE DIVISION OF FRACTIONS
ANSWERS, page 101

1) $\frac{1}{2}$

2) $\frac{1}{10}$

3) $\frac{4}{7}$

4) $\frac{2}{3}$

5) $\frac{1}{6}$

6) $\frac{6}{1}$ (6 times $1\frac{2}{3}$ = 10)

7) $\frac{1}{16}$

8) 16 (16 times $\frac{1}{2}$ is 8)

9) $6\frac{2}{3}$

10) 4

RATIO AND PROPORTION
ANSWERS, page 108

1) 24
2) 25
3) 40
4) $\frac{1}{5}$
5) $\frac{28}{9}$
6) $\frac{21}{4}$
7) $\frac{25}{2}$
8) 2
9) $\frac{5}{3}$
10) $\frac{35}{3}$
11) 47.5 miles
12) 15,625 pounds
13) 637,000 votes
14) $51\frac{1}{2}$ kg.
15) $66\frac{1}{4}$ mph
16) 22 ft
17) a) 18 miles
    b) $1\frac{7}{8}$ hours
18) 114 grams
19) 360 gallons

(Answers continued on next page)

20) 249 ounces urine
21) 52 lbs.
22) 6 gallons
23) $\frac{350}{3}$ therms or $116\frac{2}{3}$ therms
24) a) 21 inches long
b) $\frac{65}{3}$" wide × $\frac{65}{2}$" long
25) 50 strawberries
5 oz. (10 Tbsp) port
$6\frac{2}{3}$ canned figs
5 oz. (10 Tbsp) kirsch
$3\frac{2}{3}$ cups whipped cream
10 scoops vanilla ice cream
$6\frac{2}{3}$ macaroons
$3\frac{2}{3}$ Tsp. Grand Marnier
26) Only $4\frac{7}{8}$ minutes if you boil all 4 eggs together in the same pan.
27) 100 days
28) 27 days

COMPLEX FRACTIONS
ANSWERS, page 117

1) $\frac{8}{27}$
2) 20
3) $\frac{3}{16}$
4) $\frac{1}{4}$
5) 2
6) 2
7) $\frac{19}{4} = 4\frac{3}{4}$
8) $\frac{2}{3}$
9) $\frac{36}{7} = 5\frac{1}{7}$
10) $\frac{3}{5}$
11) $\frac{5}{2} = 2\frac{1}{2}$
12) $\frac{3}{2} = 1\frac{1}{2}$
13) $\frac{5}{3} = 1\frac{2}{3}$
14) $\frac{3}{7}$
15) $\frac{13}{30}$

FRACTIONS AND THE CALCULATOR
ANSWERS, page 129

1) .075
2) 2.666666
3) .0833333
4) .4166666
5) 3.8
6) .55
7) .566666
8) 7.125
9) .9166666
10) 33.333333
11) 7.625
12) .5454545
13) 4.5555555
14) 19.75
15) 89.454545
16) 66.666666
17) .02345
18) 12.5
19) .1428571
20) .2857142
21) .4285714
22) .5714285
23) .7142857
24) .8571428
25) 25.99
26) 3333.3333
27) .25
28) 25
29) 3.5625
30) .275
31) 5
32) 5.25
33) .6666666
34) .5555555
35) 4.1818181
36) 38.964844

(Answers continued on next page)

37) 4
38) 3.4
39) 1.1666666
40) 2.1666666
41) .125
42) 2.7777777
43) 1.25
44) 1.1666666
45) 7.075
46) .1333333
47) .1
48) 7.375
49) 13
50) 2.125
51) 10
52) 14.4166667
53) .125
54) .4444444
55) 0 (or .0000001 or something very small)
56) 0 (or .0000001 or something very small)
57) 6.125
58) 4.5 = 4.5
59) 5.3333333 = 5.3333333
60) 6.25 = 6.25
61) 7.2 = 7.2
62) .5 = .5
63) 1.3333333 = 1.3333333
64) 2.25 = 2.25
65) 1.5 = 1.5
66) 1.3333333 = 1.3333333
67) 1.25 = 1.25

FRACTIONS — REVIEW
ANSWERS, page 133

1) $2\frac{3}{8}$
2) $4\frac{1}{5}$
3) $7\frac{7}{10}$
4) $12\frac{1}{8}$
5) $2\frac{1}{133}$
6) $\frac{15}{2}$
7) $\frac{11}{4}$
8) $\frac{50}{3}$
9) $\frac{99}{10}$
10) $\frac{101}{2}$
11) $\frac{5}{6} = \frac{15}{18}$
12) $\frac{7}{8} = \frac{56}{64}$
13) $11\frac{3}{4} = \frac{47}{4} = \frac{188}{16}$
14) $2\frac{2}{3} = \frac{8}{3} = \frac{40}{15}$
15) $12\frac{1}{2} = \frac{25}{2} = \frac{100}{8}$
16) $\frac{9}{16}$
17) $\frac{2}{3}$
18) $\frac{5}{9}$
19) $\frac{1}{3}$
20) $\frac{8}{9}$
21) $\frac{20}{9} = 2\frac{2}{9}$
22) $\frac{2}{5}$
23) $\frac{4}{11}$
24) $\frac{3}{4}$
25) $\frac{3}{77}$
26) $\frac{5}{12}, \frac{3}{8}, \frac{1}{3}$
27) $\frac{4}{9}, \frac{44}{100}, \frac{439}{999}, \frac{43}{99}$
28) $1\frac{36}{70}, 1\frac{1}{2}, 1\frac{3}{7}, 1\frac{5}{14}$
29) $\frac{1}{2}$
30) $2\frac{11}{30}$
31) $2\frac{3}{10}$

(Answers continued on next page)

32) $1\frac{1}{9}$

33) $9\frac{1}{6}$

34) $17\frac{1}{10}$

35) $2\frac{1}{21}$

36) $3\frac{11}{12}$

37) $4\frac{5}{6}$

38) $\frac{7}{9}$

39) $\frac{2}{3}$

40) $6\frac{2}{3}$ or $\frac{20}{3}$

41) $\frac{1}{14}$

42) 15

43) $1\frac{1}{2}$ or $\frac{3}{2}$

44) $\frac{3}{2}$

45) $\frac{27}{64}$

46) $6\frac{1}{4}$ or $\frac{25}{4}$

47) $\frac{1}{6}$

48) 27 sacks

49) $\frac{11}{30}$

50) 34 m.p.g.

51) $2\frac{1}{3}$

52) 340 kwh

53) a) $\frac{1}{5}$ b) 9 students

SAMPLE EXAM #1 FOR FRACTIONS
ANSWERS, page 139

1) a) $\frac{5}{12}$ b) $\frac{1}{4}$

2) 12

3) $23\frac{1}{4}$

4) $13\frac{1}{2}$

5) $11\frac{7}{24}$

6) $15\frac{7}{8}$

7) $5\frac{3}{8}$

8) $14\frac{5}{8}$

9) $3\frac{7}{12}$

10) $3\frac{5}{6}$

11) $\frac{8}{15}$

12) $\frac{7}{10}$

13) $\frac{7}{6}$ or $1\frac{1}{6}$

14) 180

15) $\frac{7}{4}$ or $1\frac{3}{4}$

16) $\frac{2}{9}$

17) $\frac{27}{10} = 2\frac{7}{10}$

18) $\frac{7}{8}$

19) $\frac{1}{4}$

20) $\frac{27}{2} = 13\frac{1}{2}$

21) $2\frac{7}{12}$

(Answers continued on next page)

22) $151\frac{7}{8}$

23) $133\frac{5}{12}$

24) 8 boards

25) $\$312\frac{3}{8}$

SAMPLE EXAM #2 FOR FRACTIONS
ANSWERS, page 141

1) a) $\frac{8}{15}$ b) $\frac{3}{8}$

2) $10\frac{1}{5}$

3) $12\frac{7}{8}$

4) $12\frac{7}{12}$

5) $13\frac{1}{24}$

6) $11\frac{11}{24}$

7) $3\frac{1}{2}$

8) $\frac{11}{12}$

9) $2\frac{19}{20}$

10) $5\frac{1}{8}$

11) $\frac{7}{25}$

12) $\frac{7}{12}$

13) $\frac{11}{6}$

14) 165

15) $\frac{3}{2}$

16) $\frac{9}{16}$

17) 2

18) $\frac{5}{24}$

19) 15

20) $\frac{33}{10} = 3\frac{3}{10}$

21) 8 bottles

22) $\frac{4}{15}$

23) $1\frac{1}{12}$ miles each way

24) $11\frac{1}{9}$ gallons

READING DECIMALS, DECIMALS TO EXPANDED FORM, AND DECIMALS TO FRACTIONS
ANSWERS, page 149

1) 4.005
2) .345
3) 11.02
4) 1234.56
5) 0.5675
6) 1.1
7) .0005
8) 4.000001
9) 1,234,000.25
10) 999.0999
11) 9-thousandths
12) 2 and 12-hundredths
13) 159 and 5078-ten-thousandths
14) 34-hundred-thousandths
15) 798 and 99-thousandths
16) 49 and 9-tenths
17) 499-thousandths
18) 499-ten-thousandths
19) 499-hundred-thousandths
20) 25 and 6666-ten-thousandths
21) a) hundredths
 b) tenths
 c) thousand
 d) ten-thousandths
22) a) thousandths
 b) ones
 c) hundred-thousandths
 d) millionths
 e) 10-millionths
23) $\frac{1}{10} + \frac{2}{100} + \frac{5}{1000}$
24) $2(10) + 3(1) + \frac{5}{10} + \frac{5}{100} + \frac{5}{1000} + \frac{5}{10000} + \frac{5}{100000}$

(Answers continued on next page)

25) $\frac{0}{10} + \frac{0}{100} + \frac{9}{1000}$

26) $\frac{7}{10}$

27) $4(1000) + 0(100) + 0(10) + 0(1) + \frac{0}{10} + \frac{2}{100} + \frac{4}{1000}$

28) $4(1000) + 0(100) + 0(10) + 0(1) + \frac{2}{10} + \frac{4}{100} + \frac{0}{1000}$

29) $9(10) + 9(1) + \frac{0}{10} + \frac{9}{100} + \frac{9}{1000}$

30) $\frac{0}{10} + \frac{0}{100} + \frac{0}{1000} + \frac{6}{10000} + \frac{2}{100000}$

31) $\frac{8}{10} + \frac{9}{100} + \frac{0}{1000} + \frac{2}{10000}$

32) $1(1{,}000{,}000{,}000) + 0(100{,}000{,}000) + 0(10{,}000{,}000) + 0(1{,}000{,}000) + 0(100{,}000) + 0(10{,}000) + 0(1{,}000) + 0(100) + 0(10) + 0(1) + \frac{0}{10} + \frac{0}{100} + \frac{0}{1000} + \frac{1}{10000}$

33) $\frac{0}{10} + \frac{0}{100} + \frac{0}{1000} + \frac{0}{10000} + \frac{0}{100000} + \frac{0}{1000000} + \frac{0}{10000000} + \frac{0}{100000000} + \frac{5}{1000000000}$

34) $\frac{0}{10} + \frac{0}{100} + \frac{6}{1000} + \frac{0}{10000} = \frac{0}{10} + \frac{0}{100} + \frac{6}{1000}$

35) $2(10) + 5(1) + \frac{5}{10}$

36) $\frac{375}{1000} = \frac{3}{8}$

37) $\frac{5}{1000} = \frac{1}{200}$

38) $11\frac{75}{100} = 11\frac{3}{4}$

39) $245\frac{500}{1000} = 245\frac{1}{2}$

40) $245\frac{5}{10} = 245\frac{1}{2}$

41) $15\frac{75}{1000} = 15\frac{3}{40}$

42) $9\frac{625}{1000} = 9\frac{5}{8}$

43) $\frac{9}{1000}$

44) $\frac{8}{1000} = \frac{1}{125}$

45) $\frac{5}{100} = \frac{1}{20}$

46) $1987\frac{25}{100} = 1987\frac{1}{4}$

47) $10\frac{5}{10} = 10\frac{1}{2}$

48) $10\frac{5}{100} = 10\frac{1}{20}$

49) $8\frac{453}{1000}$

50) $\frac{999}{1000}$

51) $24\frac{1}{3}$

52) $66\frac{2}{3}$

EQUIVALENT DECIMALS AND ORDERING DECIMALS
ANSWERS, page 159

1) False
2) True
3) False
4) False
5) True
6) True
7) True
8) False
9) True
10) 7.9, 7.989, 7.9901
11) .80, .88, .8801, .893, .9
12) 1.0001, 1.001, 1.0011
13) .190, 1.90, 10, 10.90, 19.01, 19.10
14) 3.0003, 3.003, 3.0035, 3.03, 3.031
15) 5.7, 5.76999, 5.77, 5.7769, 5.777, 5.780, 5.79, 5.799

16) 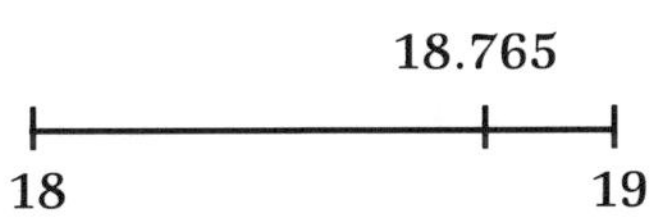

17) 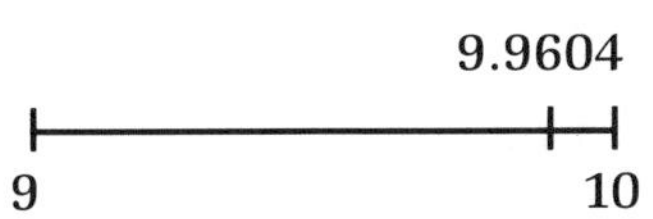

18) 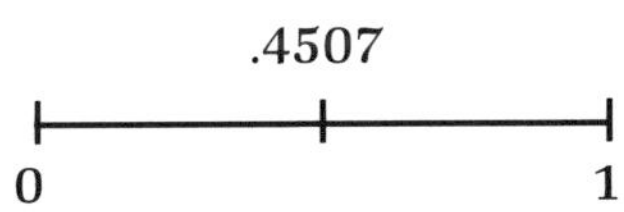

19) 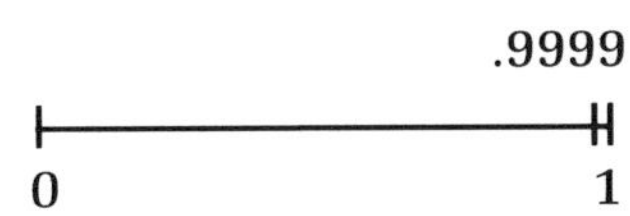

20) 

21) 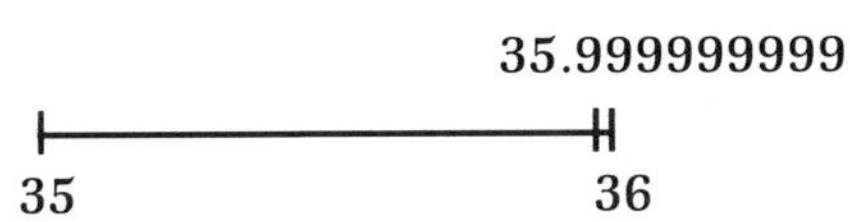

22) 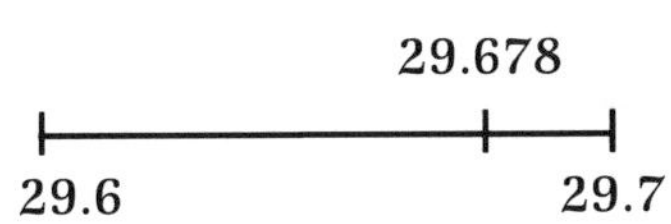

100.001

(Answers continued on next page)

23) 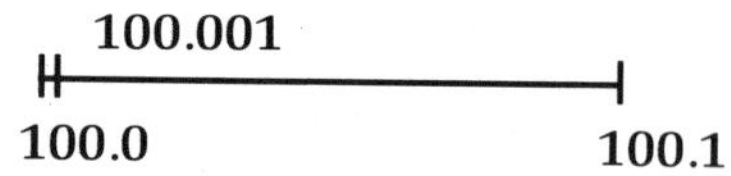

24) 

25) 

26) 

27) 

28) 

29) 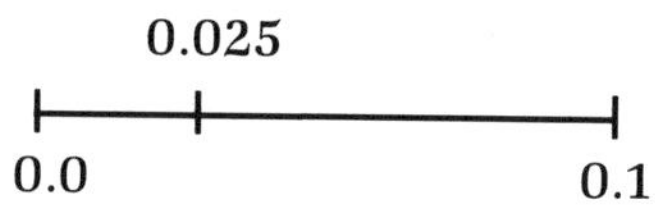

30) 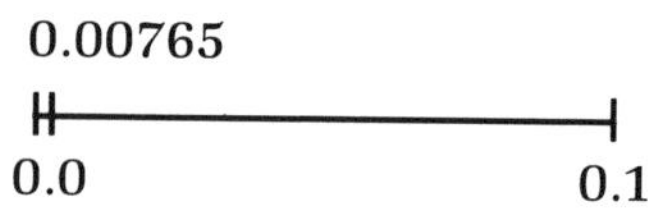

31) 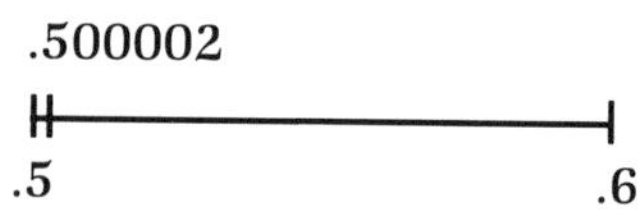

32) 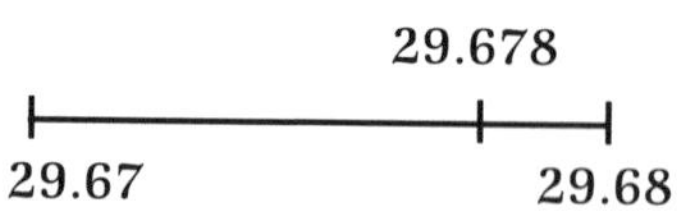

33) 

34) 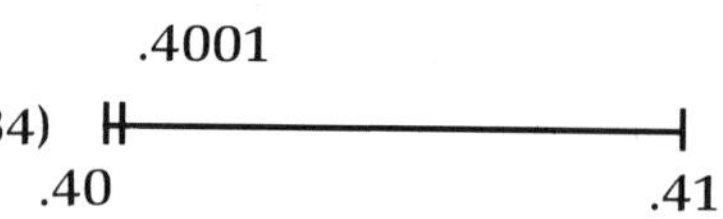

35) 

36) 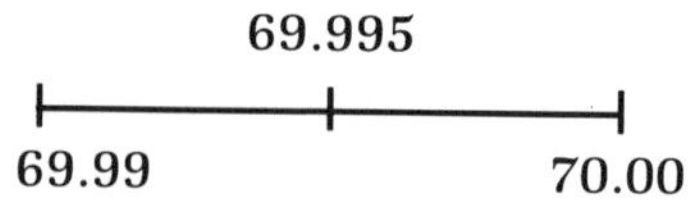

37) 

38) 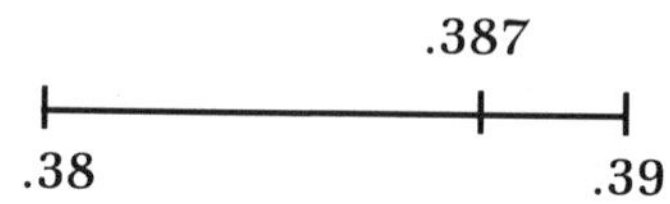

39) 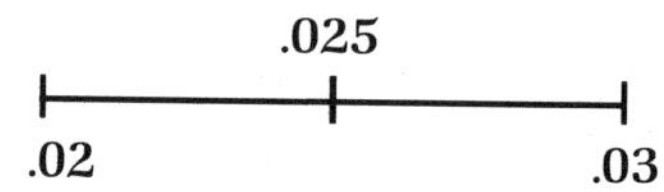

40) 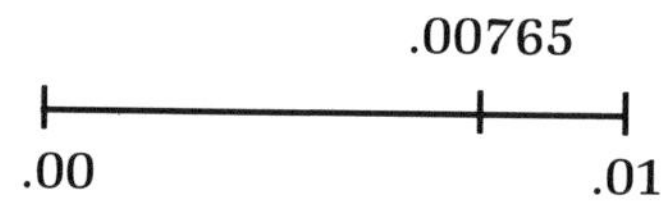

41) 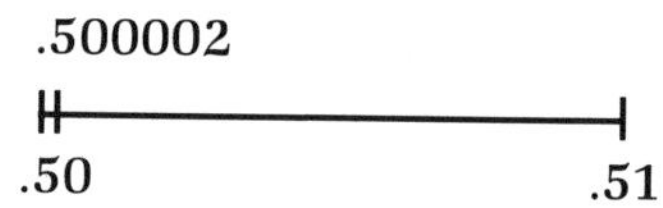

42) 

43) 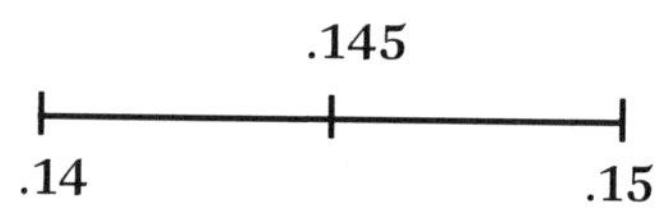

44) 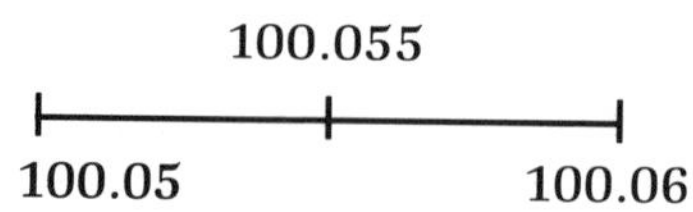

45) 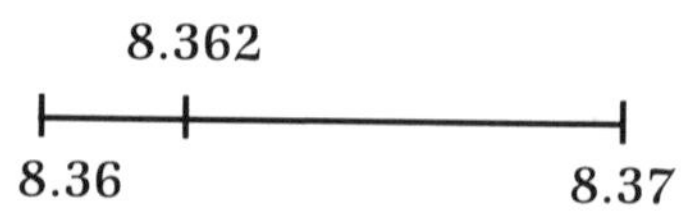

46) 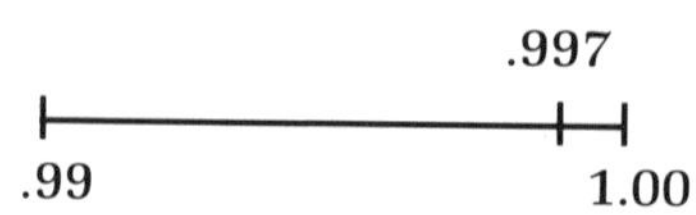

47) 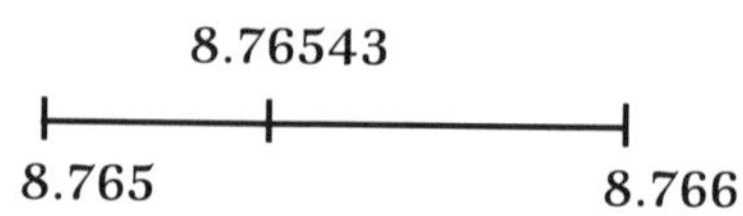

(Answers continued on next page)

48) 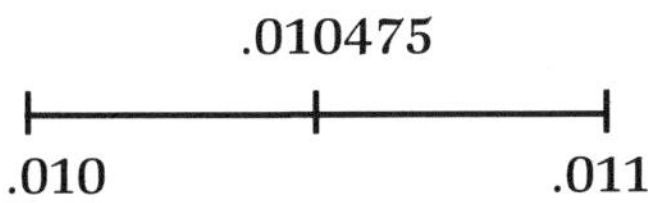

49) 

50) 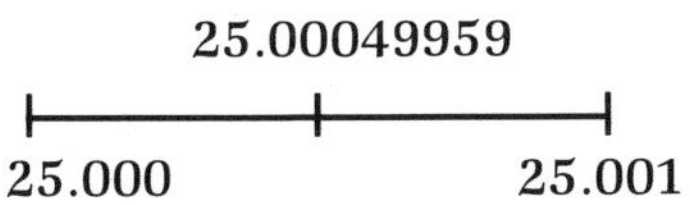

ADDITION AND SUBTRACTION OF DECIMALS
ANSWERS, page 165

1) 3.379
2) 162.8304
3) 264.888
4) 18.55
5) $1046.49
6) .0013000789
7) 5.375
8) 8.675
9) 2.0761
10) .4111
11) 126.90193
12) $4587.989
13) 11.019
14) .7928
15) 6.25
16) $25.85
17) 75.001
18) 11.2699
19) $375.46
20) 1.234

APPLICATIONS OF ADDITION AND SUBTRACTION OF DECIMALS
ANSWERS: page 167

1) $11.05
2) 1.32 ft.
3) He needs 25.95 feet
4) $2958.67
5) 4.4 kwh
6) $1097.76
7) .45
8) 51.4
9) a) 3.9 million b) 23.3 million increase
10) bottle $1.05, cup of ice $.05
11) $8.75

MULTIPLICATION AND DIVISION BY POWERS OF 10
ANSWERS, page 175

1) 3400
2) 34
3) 148,000
4) 1480
5) 148
6) 80
7) 8
8) .008
9) 125
10) 3000
11) 1250
12) 1250
13) 11,000
14) 110
15) 110
16) 17,580
17) 23.75
18) 62.5
19) 100,000
20) 1000
21) 1.1
22) .11
23) 12.4
24) .00124
25) 7.69
26) .0769
27) 7.69
28) .0769
29) .0769
30) .000769
31) .4
32) .004
33) 10
34) .001
35) .1
36) .095
37) .1625
38) .00073
39) 15.2
40) 3.89
41) $\times 100$
42) $\div 1000$
43) $\div 10{,}000$
44) $\div 100$
45) $\times 100$
46) $\times 1000$
47) $\div 100$

(Answers continued on next page)

48) ÷10
49) ÷1000
50) ×1000
51)

#	Multiply by 100	Divide by 100	Divide by 1000	Multiply by 10
24	2400	.24	.024	240
18	1800	.18	.018	180
7.5	750	.075	.0075	75
1	100	.01	.001	10
.4	40	.004	.0004	4
30	3000	.3	.03	300
6.8	680	.068	.0068	68
9	900	.09	.009	90

SCIENTIFIC NOTATION
ANSWERS, page 181

1) 1.7×10^2
2) 4.5×10^3
3) 1.02×10^3
4) 6.5×10^4
5) 2.91×10^7
6) 8.015×10^{12}
7) 1.67×10^2
8) 5×10^0
9) 2.8×10^1
10) 3.48×10^1
11) 8.05×10^5
12) 7.98×10^{14}
13) 4.9816×10^2
14) 1.523×10^2
15) 1.95×10^1
16) 7.8×10^0
17) 5.38×10^4
18) 6.74×10^2
19) 1.1101×10^8
20) 9.99×10^{21}
21) 370
22) 4650
23) 1,010,000
24) 79.8
25) 95,600
26) 500,460,000,000,000,000,000
27) 8000
28) 4000
29) 7,000,000
30) 7
31) 190
32) 416,000,000,000,000
33) 10,000,000,000
34) 2000
35) 600
36) 580,000,000,000
37) 1,530,000,000,000
38) 99,000,000,000,000,000
39) 4100
40) 7500
41) 3.2×10^{17}
42) 3.15×10^7
43) 8.69×10^5
44) 1.46×10^4
45) 1×10^5
46) 7×10^6; 1.8×10^2
47) 3×10^8
   a) 1.8×10^{10}
   b) 1.08×10^{12}
   c) 2.592×10^{13}
   d) 9.4608×10^{15}
   e) 4.068144×10^{16}

ANSWERS TO MULTIPLICATION OF DECIMALS EXERCISES
ANSWERS, page 187

1) .625
2) .612
3) .1089
4) .04096
5) 22.275
6) .000360 or .00036
7) 5.9708
8) .00512
9) 1.728
10) 16.000 or 16
11) .75
12) 10
13) .0096
14) .0441
15) .21
16) .000021
17) 4.84
18) .0016
19) 2.25
20) .00000081
21) .125
22) 30.75

(Answers continued on next page)

23) .121
24) .759
25) .0225
26) .0576
27) .00132
28) 5.1
29) .17365
30) 2.088

DIVISION OF DECIMALS
ANSWERS, page 195

1) .003
2) .006
3) 2.06
4) .15
5) .0025
6) .0114
7) .36
8) 1.5
9) .00013
10) 4
11) 2.2
12) 45000
13) 5.2
14) .018
15) 30000
16) .007
17) .36
18) $.\overline{6}$
19) .75
20) $133.\overline{3}$
21) .475
22) .4
23) .012
24) .0008
25) .00011
26) .11
27) 40.1
28) 110
29) 5
30) 231
31) 302
32) 168.018
33) 60
34) 3.2
35) .375
36) .5
37) $.\overline{428571}$
38) $.208\overline{3}$
39) .8
40) .000002
41) .04
42) 35.6
43) $66.\overline{6}$
44) $9.\overline{09}$

APPLICATION OF MULTIPLICATION AND DIVISION OF DECIMALS
ANSWERS, page 199

1) 150.5 miles per hour
2) 70 hours or 2.92 days
3) 8.75 hours
4) 13.56 grams
5) 70 inches
6) $3650.
7) 24 pencils
8) $4.11
9) 60 lbs at $384 is $6.40/lb.
10) $6.50 or 650¢
11) 56.82 kg
12) 56.52 inches
13) 830 seconds or 13 minutes, 50 seconds
14) 72 ounces at $3.99 is $.0554 per ounce versus $.0688 per ounce
15) $11.01 per pound
16) $198
17) $1152.48
18) 1200 miles
19) 191
20) $112.80

CHANGING FRACTIONS TO DECIMALS AND DECIMALS TO FRACTIONS
ANSWERS, page 207

1) .6
2) $.\overline{1}$
3) .38
4) $1.\overline{6}$
5) 2.1
6) .875
7) $.41\overline{6}$
8) $.\overline{18}$
9) $1.1\overline{6}$
10) 2.4
11) $66.\overline{6}$
12) 12.5

(Answers continued on next page)

13) .25
14) .75
15) .999
16) $.1\overline{6}$
17) .04
18) $2.\overline{36}$
19) 6.25
20) .05

21) a) $\frac{1}{7} = .\overline{142857}$

$\frac{2}{7} = .\overline{285714}$

$\frac{3}{7} = .\overline{428571}$

$\frac{4}{7} = .\overline{571428}$

$\frac{5}{7} = .\overline{714285}$

$\frac{6}{7} = .\overline{857142}$

b) They all have the digits 1, 4, 2, 8, 5, 7 that repeat in that order but starting at different points.

c) The numerator times the repeating digits of $\frac{1}{7}$ determines the digits of the other sevenths. For example: 2 × 142857 = 285714 so $\frac{2}{7} = .\overline{285714}$

22) a) $\frac{1}{11} = .\overline{09}$ $\frac{1}{111} = .\overline{009}$

b) $\frac{1}{111}$ has one more 1 in the denominator and one more 0 in the repeating form.

c) $\frac{2}{11} = .\overline{18}$ $\frac{2}{111} = .\overline{018}$

d) $\frac{2}{111}$ has one more 1 in the denominator than $\frac{2}{11}$ and its repeating part has an extra zero.

e) $\frac{5}{11} = .\overline{45}$

f) $\frac{5}{111} = .\overline{045}$

g) $\frac{5}{1111} = .\overline{0045}$

23) $\frac{75}{100} = \frac{3}{4}$

24) $\frac{125}{1000} = \frac{1}{8}$

25) $\frac{805}{1000} = \frac{161}{200}$

26) $\frac{8}{10} = \frac{4}{5}$

27) $\frac{4}{10} = \frac{2}{5}$

28) $7\frac{25}{100} = 7\frac{1}{4} = \frac{29}{4}$

29) $10\frac{35}{100} = 10\frac{7}{20} = \frac{207}{20}$

30) $2\frac{625}{10000} = 2\frac{1}{16} = \frac{33}{16}$

31) $1\frac{6}{100} = 1\frac{3}{50} = \frac{53}{50}$

32) $15\frac{2}{10} = 15\frac{1}{5} = \frac{76}{5}$

33) $100\frac{5}{100} = 100\frac{1}{20} = \frac{2001}{20}$

34) $\frac{2}{3}$

35) $\frac{5}{6}$

36) $\frac{2}{9}$

37) $\frac{25}{33}$

38) $\frac{37}{300}$

39) $\frac{61}{495}$

40) $\frac{41}{333}$

DECIMALS — REVIEW
ANSWERS, page 213

1) 15.023
2) .0101
3) 2456.39
4) 2.000002
5) 2 and 18-thousandths
6) 1987 and 45-ten thousandths
7) 6 and 6-hundred thousandths

8) $8\frac{5}{8} = \frac{69}{8}$

9) $14\frac{1}{40} = \frac{561}{40}$

10) $37\frac{1}{3} = \frac{112}{3}$

11) $298\frac{9}{10} = \frac{2989}{10}$

(Answers continued on next page)

12) $1987\frac{3}{400} = \frac{794803}{400}$

13) 2.088, 2.089, 2.8, 2.88, 2.88009, 2.8801
14) 1.749, 1.75, 1.75009, 1.7501, 1.755
15) 1.0001, 1.00011, 1.001
16) 7.7204
17) 214.9914
18) 20.22
19) 44.2341
20) 72.5925
21) 2.45×10^8
22) 1.798×10^1
23) 8×10^2
24) 4.044×10^{13}
25) 9.23×10^0
26) 1×10^{13}
27) .0684
28) 38.22
29) .297
30) 187.5
31) .04
32) .018
33) 170
34) 11
35) $2.\overline{333}$
36) 1.8
37) .342
38) .0135
39) 5.76
40) .000936
41) .00007
42) .39
43) 203.05
44) 3.8
45) $.\overline{222}$
46) 2.125
47) 2.25
48) $.58\overline{3}$
49) .65
50) $.208\overline{3}$

51) $\frac{149}{200}$

52) $13\frac{7}{2000} = \frac{26007}{2000}$

53) $2\frac{7}{8} = \frac{23}{8}$

54) $19\frac{1}{25} = \frac{476}{25}$

55) $\frac{7}{9}$

56) $\frac{17}{30}$

57) No. He needs .4 feet more than he needs.
58) 85.2 ml.
59) 500
60) $54.00

SAMPLE DECIMALS EXAM
ANSWERS, page 221

1) 1908.71363
2) 161.37807
3) 5.18
4) 493.017
5) 130.68
6) 117.260
7) $.\overline{333}$
8) .012
9) 2.5
10) .8
11) a) .00004512
    b) 9.2630
12) a).19 b) .875 c) $.\overline{666}$
13) a) $\frac{3}{8}$ b) $\frac{1}{40}$ c) $\frac{3}{2000}$
14) .661, .65, .649001, .649
15) Needs 9.95 feet more fencing
16) a) 87600 b) .00285 c) .09 d) .06
17) a) 16.2146 b) 3.5858625 c) 43902.5
    d) 12.4 e) $1.\overline{3636}$ f) 2.223 g) 5.5696
    h) 8 i) 17794.4 j) .3125 k) 66.666
    l) $1,234,575.23 m) $2.708\overline{3}$

PERCENTS — MEANING & DECIMAL FORM
ANSWERS, page 227

1) .19
2) .35
3) .76
4) .09
5) .45
6) 2.15
7) 1.32
8) 4.5
9) 1.55
10) .98

(Answers continued on next page)

11) .0025
12) .125
13) .7725
14) .916
15) .003
16) .175
17) .49375
18) .6525
19) .065
20) .0075
21) 63%
22) 52%
23) 7%
24) 16%
25) 48%
26) 93%
27) 80%
28) 50%
29) 3%
30) .25% or $\frac{1}{4}$%
31) 7.5% or $7\frac{1}{2}$%
32) 204%
33) 225%
34) 2.8% or $2\frac{4}{5}$%
35) $33\frac{1}{3}$%
36) 1.25% or $1\frac{1}{4}$%
37) 18.75% or $18\frac{3}{4}$%
38) $6\frac{2}{3}$%
39) $15\frac{1}{2}$%
40) 300%

USE OF PERCENTS
ANSWERS, page 235

1) 22
2) $25.50
3) 37.5
4) 20%
5) 600%
6) $\frac{1}{2}$% or .5%
7) 4.4
8) 800
9) 30
10) .4
11) 3
12) 410
13) 100%
14) 37%
15) 21
16) 80%
17) 125%
18) 80
19) 28
20) $16\frac{2}{3}$% or $16.\overline{6}$%
21) $37\frac{1}{2}$% or $37.\overline{5}$%
22) $.65
23) 400%
24) 135
25) 42

APPLICATION OF PERCENTS
ANSWERS, page 245

1) 70%
2) 63 taste buds
3) $5.32
4) $15,360.
5) a) 8 billion acres
   b) 3,440,000,000 acres
6) a) 48.9%
   b) 52.4%
7) a) 48.1%
   b) 11.1%
8) 2.7%
9) .0294% or approximately .03%
10) 780.8 milliliters of nitrogen
11) 90.7% or 90 ⅔%
12) $6285.
13) $19.48
14) 12½%
15) 148,800
16) 20.4%
17) $360.
18) 7%
19) 60%
20) $\frac{62}{311}$ = 19.9% and *is* within regulations
21) 75%
22) 2340
23) $5310
24) 574,070,000

(Answers continued on next page)

25) a) 2.73% or ≈ 2.7%
b) 1.06% or ≈ 1.1%
c) 1.67% or ≈ 1.7%
26) 25%
27) 392

PERCENTS AND THE CALCULATOR
ANSWERS, page 255

1) .47
2) .36
3) .04
4) .26
5) .52
6) 1.25
7) 2.35
8) .125
9) $.\overline{333}$
10) $.\overline{666}$
11) .60
12) .0025
13) .015
14) .0275
15) .935
16) .8725
17) .005
18) .000025
19) .0108
20) .32875
21) 38%
22) 54%
23) 78%
24) 80%
25) 210%
26) 145%
27) .25%
28) 3.75%
29) 30%
30) 15%
31) 62.5%
32) 300%
33) $4\frac{1}{2}\% = 4.5\%$
34) .2%
35) .75%
36) 24
37) 80%
38) 60
39) 21
40) $66.\overline{6}\%$ or $66\frac{2}{3}\%$
41) 54
42) 250%
43) .6
44) 20

PERCENTS — REVIEW
ANSWERS, page 257

1) .37
2) .17
3) .59
4) .06
5) .35
6) .0725
7) $.\overline{66}$
8) 2.05
9) 1.01
10) .00375
11) 79%
12) 3%
13) 16%
14) 40%
15) 142%
16) $66\frac{2}{3}\%$
17) $3\frac{3}{4}\%$ or 3.75%
18) $24\frac{3}{8}\%$ or 24.375%
19) $\frac{1}{2}\%$ or .5%
20) 400%

(Answers continued on next page)

21)

Decimal Form	Percent Form	Fraction Form
$.\overline{333}$	$33\frac{1}{3}\%$	$\frac{1}{3}$
.75	75%	$\frac{3}{4}$
.0625	$6\frac{1}{4}\%$	$\frac{1}{16}$
.4 or .40	40%	$\frac{2}{5}$
$1.\overline{33}$	$133\frac{1}{3}\%$	$\frac{4}{3}$
.125	$12\frac{1}{2}\%$	$\frac{1}{8}$
.375	$37\frac{1}{2}\%$	$\frac{3}{8}$
.48	48%	$\frac{12}{25}$
.05	5%	$\frac{1}{20}$
$.1\overline{66}$	$16\frac{2}{3}\%$ or $16.\overline{6}\%$	$\frac{1}{6}$
.0275	$2\frac{3}{4}\%$	$\frac{11}{400}$
2.3	230%	$2\frac{3}{10} = \frac{23}{10}$

22) 121.5
23) $16\frac{2}{3}\%$
24) 162
25) 45
26) 50
27) $333\frac{1}{3}\%$
28) 45
29) 125%
30) 56
31) $400
32) 25%
33) 18,000
34) $300
35) 40%

SAMPLE PERCENTS EXAM #1
ANSWERS, page 263

I. 1) .57
   2) 3.52
   3) .0075
   4) .006

II. 1) 65%
   2) 4.35%
   3) 436%
   4) 29.875% or $29\frac{7}{8}\%$

III. 1) 15
   2) 12.75
   3) 30%
   4) 6.25% or $6\frac{1}{4}\%$
   5) 68
   6) 400

IV. 1) $16.22
   2) $1080
   3) 85%
   4) $180
   5) 20%

BASICS OF THE METRIC SYSTEM: LENGTH
ANSWERS, page 269

1) 18 m.
2) 45 km.
3) 786 cm.
4) 1700 m.
5) 80,000 m.
6) 560 m.
7) 4,321,000 mm.
8) 10,000,000 mm.
9) 4828.03 km.
10) 18.5 dam.
11) .53 cm.
12) .028 km.
13) 284 km.
14) 240 cm.
15) 2800 mm.
16) 82.5 cm
17) 25 m.
18) 15 dm.
19) 146.3 dam.
20) 5860 mm.

BASICS OF THE METRIC SYSTEM: WEIGHT
ANSWERS, page 273

1) 4800 mg.
2) 850 kg.
3) 19.81 g.
4) 3,000,000 mg.
5) .145 kg.
6) .0379 kg.
7) 180 hg.
8) 1,800,000 cg.
9) 18 kg.
10) 2500 g.
11) 3.7 g.
12) 500 g.
13) .0195 kg.
14) .45 kg.
15) 17.7 dag.
16) .001 kg.
17) 15 hg.
18) 4750 mg.
19) 1970 mg.
20) 5.463 kg.

BASICS OF THE METRIC SYSTEM:
VOLUME AND CAPACITY
ANSWERS, page 279

1) 4500 ml.
2) 18 kl.
3) 12.5 cl.
4) 89 l.
5) 875 ml.
6) 500 ml.
7) 15 l.
8) 4.85 hl.
9) 2000 cl.
10) 125 ml.
11) .000001 m^3.
12) 5000 dm^3.
13) .001 m^3.
14) 500,000 cm^3.
15) 250 mm^3.

ADDITIONAL METRIC SYSTEM
ANSWERS, page 281

1) C
2) B
3) B
4) C
5) C
6) C
7) C
8) A
9) B
10) C
11) C
12) B
13) A
14) A
15) A
16) half (or .6) a mile)
17) 10%
18) half
19) quart
20) 10%
21) $\frac{1}{5}$

(Answers continued on next page)

22) 3 m. = 3000 mm.
3 km. = 3000 m.
3 m. = .003 km.
91.44 m. = 9144 cm.
150 cm. = 1.5 m.
1500 mm. = 150 cm.
1500 mm. = 1.5 m.
3500 mm. = .0035 km.
23) 400 m.
24) 225 cm.
25) 35 mm.
26) 7 dm.
27) .45 kg.
28) 750 m^3.
29) 10,000 kg.
30) .7 g.
31) each answer varies
32) each answer varies
33) $187\frac{1}{2}$
34) 67 Tbsp.
35) a) hectare = 100 ares = 100(100m^2.) = 100 × 100 m^2 = 100 m. × 100 m. = 10,000 m^2.
b) $\frac{1000 \text{ m.} \times 1000 \text{ m.}}{100 \text{ m.} \times 100 \text{ m.}} = 10 \times 10 = 100$ hectares

---

CONVERSIONS BETWEEN THE ENGLISH AND METRIC SYSTEM
ANSWERS, page 291

1) a) 4.1666667 or $4.1\overline{6}$ or $4\frac{1}{6}$ cups
b) 3.3333333 or $3.\overline{3}$ or $3\frac{1}{3}$ cups
2) a) 164 feet
b) $109.\overline{3}$ paces or $109\frac{1}{3}$ paces
3) $\frac{2907.8 \text{ miles}}{60 \text{ mph}} = 48.4\overline{63}$ hours
4) hectare = 100 m. × 100 m. =
100 (3.28 ft) × 100 (3.28 ft) = 107,584 sq. ft.

$$\text{acre} = \frac{\text{hectare}}{2.5} = \frac{197{,}484 \text{ sq. ft.}}{2.5} = 43{,}034 \text{ sq. ft.}$$

Actual measurement of acre = 43,542 sq. feet (discrepancy due to round-off in factor). If we had used 2.47 instead of 2.5 we would be closer to actual measurement.
5) 101 ft. 8 inches higher
6) 12 ounces
7) meter is longer by 3 inches
8) $6.588 \times 10^{21} \times 2000 = 13.176 \times 10^{24}$ lbs. = 5981.904×10^{24} grams = 598.1904×10^{21} kg. The answer is approximately 5.982×10^{21} metric tons.
9) a) 9.75 inches (almost 10 inches)
b) 0.975 inches (almost 1 inch)
c) 3.06 inches
10) female — 2 lbs. 12 oz.
male — 3 lbs. 6 oz.
11) 485.78 kg.

$$(1070 \text{ lb.} \times \frac{454 \text{ g.}}{1 \text{ lb.}} \times \frac{1 \text{ kg.}}{1000 \text{ g.}} = 485.78 \text{ kg.})$$

12) 40,250,000 meters (use 1 km. = 1.61 miles)
13) 1128.32 ft/sec
1266.08 ft/sec
14) Approx 4921 beetles
15) 4540 kg.
16) 3.3 cups

$$750 \text{ g.} \times \frac{1 \text{ kg.}}{1000 \text{ g.}} \times \frac{2.2 \text{ lb.}}{1 \text{ kg.}} \times \frac{16 \text{ oz.}}{1 \text{ lb.}} \times \frac{1 \text{ cup}}{8 \text{ oz.}}$$

17) a) 85 oz.
b) 51 oz.
c) 17 oz.
18) 328 feet

TABLE OF PRIME NUMBERS
(between 1 and 1,000)

2	191	433	701
3	193	439	709
5	197	443	719
7	199	449	727
		457	733
11	211	461	739
13	223	463	743
17	227	467	751
19	229	479	757
23	233	487	761
29	239	491	769
31	241	499	773
37	251		787
41	257	503	797
43	263	509	809
47	269	521	811
53	271	523	821
59	277	541	823
61	281	547	827
67	283	557	829
71	293	563	839
73		569	853
79	307	571	857
83	311	577	859
89	313	587	863
97	317	593	877
	331	599	881
101	337		883
103	347	601	887
107	349	607	
109	353	613	907
113	359	617	911
127	367	619	919
131	373	631	929
137	379	641	937
139	383	643	941
149	389	647	947
151	397	653	853
157		659	967
163	401	661	971
167	409	673	977
173	419	677	983
179	421	683	991
181	431	691	997